T0210658

Advanced Concepts and Architectures for Plasma-Enabled Material Processing

Synthesis Lectures on Emerging Engineering Technologies

Editor
Kris Iniewski, *CMOS ET*

© Springer Nature Switzerland AG 2022
Reprint of original edition © Morgan & Claypool 2020

All rights reserved. No part of this publication may be reproduced, stored in a retrieval system, or transmitted in any form or by any means—electronic, mechanical, photocopy, recording, or any other except for brief quotations in printed reviews, without the prior permission of the publisher.

Advanced Concepts and Architectures for Plasma-Enabled Material Processing
Oleg O. Baranov, Igor Levchenko, Shuyan Xu, and Kateryna Bazaka

ISBN: 978-3-031-00907-5   paperback
ISBN: 978-3-031-02035-3   ebook
ISBN: 978-3-031-00141-3   hardcover

DOI 10.1007/978-3-031-02035-3

A Publication in the Springer series
SYNTHESIS LECTURES ON EMERGING ENGINEERING TECHNOLOGIES

Lecture #11
Series Editor: Kris Iniewski, CMOS ET
Series ISSN
Print 2381-1412   Electronic 2381-1439

Advanced Concepts and Architectures for Plasma-Enabled Material Processing

Oleg O. Baranov
National Aerospace University, Kharkiv, Ukraine, and Jožef Stefan Institute, Ljubljana, Slovenia, EU

Igor Levchenko
Nanyang Technological University, Singapore, and Queensland University of Technology, Brisbane, Australia

Shuyan Xu
Nanyang Technological University, Singapore

Kateryna Bazaka
The Australian National University, ACT, Australia, Queensland University of Technology, Brisbane, Australia, and Nanyang Technological University, Singapore

SYNTHESIS LECTURES ON EMERGING ENGINEERING TECHNOLOGIES #11

ABSTRACT

Plasma-based techniques are widely and successfully used across the field of materials processing, advanced nanosynthesis, and nanofabrication. The diversity of currently available processing architectures based on or enhanced by the use of plasmas is vast, and one can easily get lost in the opportunities presented by each of these configurations. This mini-book provides a concise outline of the most important concepts and architectures in plasma-assisted processing of materials, helping the reader navigate through the fundamentals of plasma system selection and optimization. Architectures discussed in this book range from the relatively simple, user-friendly types of plasmas produced using direct current, radio-frequency, microwave, and arc systems, to more sophisticated advanced systems based on incorporating and external substrate architectures, and complex control mechanisms of configured magnetic fields and distributed plasma sources.

KEYWORDS

plasma, material processing, nanosynthesis, nanofabrication, magnetic fields, direct current systems

Contents

Introduction

Material processing has a long history spanning millennia, dating back to the use of stone tools by early humans during The Early Stone Age. Since then, we have come a long way in our use of tools for material synthesis and processing. The tools of today offer an unprecedented level of sophistication and precision, allowing us to put together material architectures with nanoscale control over their structure and properties. The latter was made possible by us learning how to produce and control the assembly of increasingly small elements from which these architectures can be built, and harnessing the control over the energy and movement of particles that we could use as tools to shape and remove the material with nanoscale and atomic-scale precision.

Shaping of hard materials using fast-moving soft matter has started with the use of high-pressure (hundreds of atm) water jets [1] and hot gas jets produced by the reactors similar to rocket engines to cut and process the surface of metals and hard stones [2]. Fundamentally different in their use of a softer matter to process a harder substance, these approaches have brought to life some unique capabilities such as much higher treatment rates and in some cases lower price, especially for the treatment of very hard materials such as stones and hard, quenched metals and alloys. Rather than relying on the properties of the material from which the tool was made, e.g., hardness of a diamond or tungsten alloy, it was the energy put into the system that determined the treatment outcome, making the process far more flexible and easy to optimize.

Nevertheless, the liquid- and gas-based material treatment techniques have fundamental limitations, namely the energy that may be transferred to the process media using conventional pressure and temperature means. Creating liquid jets with pressures exceeding hundreds of atm is not trivial due to the need for tanks that could withstand exceedingly high pressures, and associated considerable energy requirements and safety implications of using liquids and gases at high pressures. More important, however, is the limited control over the interactions that take place between the accelerated particles and the materials they contact, particularly for applications that require precise yet efficient excavation and/or delivery of material to the surface without affecting the material bulk.

Unlike liquid or gas, *plasma* offers a pathway for increasing both the efficiency and resolution of material processing. As a forth state of matter, plasma consists of energetic particles that are ionized, which means that their acceleration and directed motion can be effectively controlled by *electric and magnetic fields*. This allows plasma to transfer extremely high power densities to the surface while controlling the composition, shape, temperature, density, and many other parameters of the particle flux [3].

The mechanisms by which these ionized particles are generated and controlled and the types of materials they can synthesize or process is dependent on the particulars of the plasma system. Currently, many types of plasma systems exist, including direct current (DC), radio-frequency (RF), capacitively coupled plasma (CCP), inductively coupled plasma (ICP), microwave (MW), and arc and atmospheric plasmas to name but a few. Importantly, each system features a different set of plasma parameters, jointly covering a very wide spectrum of power densities (from space debris-removing concentrated jets to gentle glow discharges for the surface processing of finely resolved organic films), resolution, and material types, and allowing the user to select the most suitable tool for their fabrication and processing needs [4].

HOW TO USE PLASMA FOR TECHNOLOGY

Usefulness of plasma as a powerful tool in material synthesis and processing has been demonstrated by the decades of scientific research and industrial use. Rich diversity of species, the widest range of possible densities and particle energies, ability to exhibit a collective behavior—these are some of the reasons for why processing tools based on this state of matter is in high demand [5, 6]. As a quasi-neutral mixture of charged particles, plasma consists of ions, electrons, excited neutrals, radicals, and neutrals, and each of these constituents can contribute some unique properties to the material processing. Ions are the carriers of the chemical properties, and their high energy can be controlled by the use of quite simple technical solutions. Electrons transfer their energy to the treated material without changing the chemical composition of the latter; for this reason, the electrons in plasma represent a valuable means for the material heating without changing its chemical composition. It is worth noting that while the density of the plasma electrons may be controlled, their energy is difficult to change across a wide range. Energy delivered by excited neutrals and radicals exceed the energy of atoms and molecules in their main state yet is less that the energy of the ions; thus, the effect of these excited species ranges from the weak action by the neutrals to significant effects caused by the ions. Very high chemical activity of the radicals makes them suitable building blocks for the synthesis of materials with controlled chemical properties, while the excited atoms and molecules may act as a tool to control the reaction rates in plasma or in a treated material. By drawing an analogy with a building yard, the rational application of the properties of the plasma species in the process of material synthesis, assembly and modification may proceed as follows.

1. *Excavation and ramming*—Ion bombardment is used to remove excessive or polluted layer, to implant doping element, and to generate a large number of defects. This produces a surface that is "prepared" or functionalized, and the surface energy is adjusted to the desired level. Usually the energy of the ions is from tens of eV to tens of keV. At such energy levels, the creation of chemical compositions that violates the thermal solution limits are possible, i.e., we have the advantages of the surface-plasma non-equilibrium chemistry. The power supplied to the surface may be controlled through the control of the ion energy and current density (or plasma density) in order to prevent the surface from melting and losing the geometrical pattern, undesired changes of morphology and chemical composition due to

collapse of the crystalline structure, and excessive diffusion and segregation of the chemical components.

2. *Foundation assembly*—Various routes may be used such as:

   (a) modification of the existing pattern through the mild ion bombardment that stimulates ion-induced diffusion and segregation. As a result of this treatment, the surface relief becomes of a regular pattern with the specified distribution of the surface energy (i.e., generation of the adsorption nodes). If the change in the chemical composition of the substrate is not desired, the electron bombardment may be applied to selectively induce the heating of the surface without changing the chemical composition; and

   (b) external delivery of the adjunct material by the deposition of vapors which deliver the neutrals as the main building blocks; or deposition of clusters and droplets, where the groups of neutrals in a liquid or solid states may act as the seeds of a new phase;

3. *Building*—Delivery of excited neutrals, radicals, and neutrals that are deposited over the foundation pattern, thus forming the base structure of the final material;

4. *Finishing*—Removal of excess material to resolve the final structure of the construct. During this stage, ion bombardment may be combined with chemical treatment caused by the chemically active neutrals, where the role of the latter is to break the bonds within the treated material and enhance the rate of material removal due to ion bombardment.

It stands to reason that the analogy does not reflect the complexity of the real processes in the plasma reactors. In the following chapters, we will attempt to describe some aspects of the available processes and mechanisms in plasma by presenting a description and analysis of the advanced concepts and architectures frequently used for plasma-enabled materials processing.

REFERENCES

[1] Miller, R. K. 1989. Waterjet cutting, *Industrial Robot Handboook: VNR Competitive Manufacturing Series*, Spring, Boston, MA. DOI: 10.1007/978-1-4684-6608-9. 1

[2] Krajcarz, D. 2014. Comparison metal water jet cutting with laser and plasma cutting, *Proc. Eng.*, 69:838–843. DOI: 10.1016/j.proeng.2014.03.061. 1

[3] Levchenko, I., Xu, S., Mazouffre, S., Lev, D., Pedrini, D., Goebel, D., Garrigues, L., Taccogna, F., and Bazaka, K. 2020. Perspectives, frontiers, and new horizons for plasma-based space electric propulsion, *Phys. Plasmas*, 27:020601. DOI: 10.1063/1.5109141. 1

[4] Lorello, L., Levchenko, I., Bazaka, K., Keidar, M., Xu, L., Huang, S., Lim, J. W. M., Xu, S. 2017. Hall thrusters with permanent magnets: Current solutions and perspectives, *IEEE Trans. Plasma Sci.*, 46:239–251. DOI: 10.1109/TPS.2017.2772332 2

[5] Levchenko, I., Bazaka, K., Baranov, O., Sankaran, R. M., Nomine, A., Belmonte, T., and Xu, S. 2018. Lightning under water: Diverse reactive environments and evidence of synergistic effects for material treatment and activation, *Appl. Phys. Rev.*, 5(2):021103. DOI: 10.1063/1.5024865. 2

[6] Bazaka, K., Levchenko, I., Lim, J. W. M., Baranov, O, Corbella, C., Xu, S., and Keidar, M. 2019. MoS2-based nanostructures: synthesis and applications in medicine, *J. Phys. D, Appl. Phys.*, 52(18):183001. DOI: 10.1088/1361-6463/ab03b3. 2

CHAPTER 1

Technological Plasmas and Typical Schematics

ABSTRACT

This chapter introduces technological plasmas as a powerful means for material synthesis and processing, and describes general features of various systems typically used for the generation and control of such plasmas. Direct current, radio-frequency, capacitively, and inductively coupled plasmas, as well as microwave and arc plasma systems, are briefly characterized and illustrated.

1.1 DIRECT CURRENT PLASMA

Powering two electrodes by a DC source is the simplest way to ignite plasma. In this configuration, the electric field is generated between the electrodes, and the electrons which are always present in the inter-electrode gap are accelerated in the direction of the positively charged electrode (the anode). In the process of moving toward the electrode, the electrons collide with the neutrals and ionize them, thus producing more electrons and ions in the ionization avalanche. The ions are accelerated by the electric field toward the negatively charged electrode (the cathode), and produce secondary electrons as a result of the collision with the cathode surface. In turn, these electrons produce new ions, and under certain conditions, the glow discharge is ignited [1–3]. For the discharge to become self-sustaining, the electron avalanche should create ions in quantities sufficient for the generation of secondary electrons [4]:

$$A\frac{d}{\lambda_{iz}}\exp\left(-\frac{\varepsilon_{iz}}{E\lambda_{iz}}\right) = \ln\left(1 + \frac{1}{\gamma_{see}}\right), \tag{1.1}$$

where d is the length of the discharge gap, E is the electric field, λ_{iz} is the mean free path of electron for inelastic collisions, ε_{iz} is the energy of ionization, and γ_{see} is a coefficient of the secondary electron emission, and A is a constant.

Obviously, the electrons will lose the energy acquired in the discharge gap along their path to the anode; thus, the necessity of maintaining the relatively high gas pressure in the gap is implied. However, due to the necessity to maintain a certain level of chemical purity, production yield and other considerations, the typical pressure of the glow discharge (above 100 Pa) is considered too high for certain types of applications, and various other techniques are developed to overcome the pressure limitation [5–8].

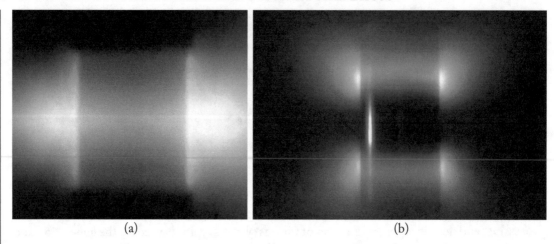

(a)   (b)

Figure 1.1: Photograph of a DC discharge (a) and its transition to a magnetron discharge (b) once the arched magnetic field is applied. Photos by the authors.

To increase the residence time of the electrons in the discharge gap, the magnetic field of a few tens of milliteslas (mT) may be applied. Once such a field is generated, the path of the charged particles becomes helical; however, the light electrons rotate with the radius that is much smaller than the extension of the discharge gap, while the heavy ions are almost not affected by the magnetic field in the gap due to their large gyro-radius. As a result, the electrons are "magnetized" while the ions are not, and in this state the transition from the conventional to the magnetron discharge occurs. An example of the transition is shown in Figure 1.1, when two coils are mounted within the volume of the cylindrical cathode. If the coils are not powered, the DC glow occurs, as shown in Figure 1.1a; once the coils are powered by the electric currents of opposite direction, two plasma tori of the magnetron discharge are ignited, as shown in Figure 1.1b.

The productivity of the plasma setups depends greatly on the number density of the plasma particles, and for the first approach the density is determined by the particle balance in the volume occupied by the plasma [9]:

$$\frac{dn}{dt} = n n_a K_{iz} - 1.72 \frac{n u_B}{l \left[3 + \frac{l}{2\lambda_i} + \frac{T_i}{5T_e} \left(\frac{l}{\lambda_i} \right)^2 \right]^{1/2}}, \tag{1.2}$$

where n is the plasma density (and also the density of the ions and electrons since plasma is considered quasineutral); u_B is the Bohm velocity; $u_B = (kT_e/M)^{1/2}$, M is the ion mass; n_a is the density of the background gas pressure; T_e and T_i are the electron and ion temperatures, respectively; K_{iz} is the temperature-dependent term that describes the ionization rate; l is the distance between the electrodes; and λ_i is the ion-neutral mean free path.

When the magnetic field is applied, the motion of the plasma electrons is described by the equation [10]:

$$mn\frac{d\vec{v}}{dt} = -en\left(\vec{E} + \vec{v} \times \vec{B}\right) - \nabla p - mn\nu\vec{v}, \tag{1.3}$$

where m, $\vec{v}$, and e are the mass, electron velocity, and charge, respectively; $\vec{E}$ is the electric field, $\vec{B}$ is the magnetic field, n is the electron density, ∇p is the pressure gradient; and ν is a frequency at which electrons collide with neutral species.

In a stationary mode, the drift velocity of the electron across the magnetic field is defined by the magnetic field B:

$$\vec{v}_{\perp} = \left[1 + \left(\frac{eB}{m\nu}\right)^2\right]^{-1}\left(-\mu\vec{E} - D\frac{\nabla n}{n} + \left(\frac{e}{m\nu}\right)^2\left[\left(\vec{E} \times \vec{B}\right) + \frac{\nabla p \times \vec{B}}{en}\right]\right), \tag{1.4}$$

where μ is the electron mobility and D is the diffusion coefficient when the magnetic field is absent.

Here, the following parameters of the motion are introduced: $\vec{v}_E = \frac{\vec{E} \times \vec{B}}{B^2}$ is the electrical drift velocity; $\vec{v}_D = \frac{\nabla p \times \vec{B}}{enB^2}$ is diamagnetic drift velocity; and $\omega_{ce} = eB/m$ describes the cyclotron frequency of electrons.

The behavior of real world plasmas under the action of magnetic fields is far more complex, and various assumptions are generally used, such as collisions of electrons with (i) the fluctuation of the electrical field (Bohm conductivity), (ii) the reactor walls that confine the plasma (near-wall conductivity), or (iii) the charged particles neutralized on the reactor walls [11].

The simplest configuration to explore the influence of the magnetic field on plasma can be created by using just two electromagnetic coils placed at some distance of each other. When they are powered by the electric current flowing in the same direction, a *magnetic bottle* is generated, thus illustrating the simplest way to isolate the reactor walls from the plasma. The configuration of the magnetic field near the powered coils is called "*magnetic mirror.*" Another form of the magnetic mirror ("*magnetic cusp*") can be obtained between the coils by powering them with the opposite-flowing currents [12].

When describing the action of the magnetic fields to the plasma confinement, the minimum B_0 and maximum B_m of the magnetic field in the trap determine the possibility of "leaking" the particle through the mirrors: only the particles with the velocities directed at an angle $\theta < \theta_m$ along the magnetic field line can leave the magnetic trap [10]:

$$\frac{B_0}{B_m} = \sin^2\theta_m. \tag{1.5}$$

The principle of the magnetic confinement forms the foundation of the electromagnetic focusing and guiding of the plasma in the technological setups, where a region of plasma gen-

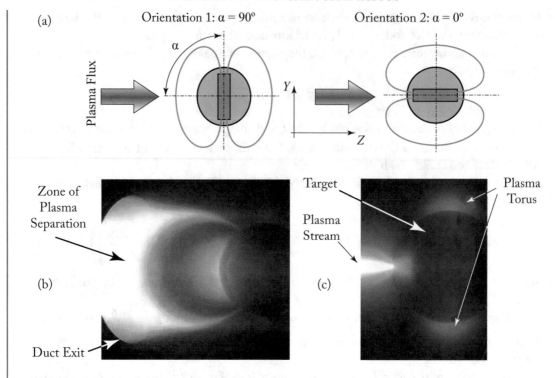

Figure 1.2: Magnetic control of an arc discharge: schematic of the magnetic field orientation relative to the plasma flux (a), photograph of the plasma structure with orientation 1, $\alpha = 90°$ (b), and orientation 2, $\alpha = 0°$ (c). Reprinted with permission from Levchenko et al., *IEEE Trans. Plasma Sci.*, 32:2139, 2004. Copyright 2004 IEEE.

eration is remote from a surface treated by the plasma. For that, the magnetic field is used to magnetize the plasma electrons and prevent them from expanding in certain directions yet allowing them to propagate freely in other directions. Plasma ions cannot move separately from the electrons due to the electric field generated at the separation, thus, the ions should follow the electrons. A number of experiments have been conducted to apply this principle in different plasma setups.

For example, plasma from the vacuum arc source was guided by Levchenko et al. [13] by means of two orientations of the treated surface (substrate) relative to the arc plasma source. Figure 1.2a shows both orientations.

For the first orientation, the magnetic field generated by a coil mounted inside of the spherical substrate is directed orthogonally with respect to the magnetic field generated by a coil wound around the plasma source. As such, the magnetic field is also perpendicular to the expansion of the plasma jet from the arc source; the configuration is similar to the interaction of the Earth's magnetic field with the solar wind. For another orientation, the magnetic field of the substrate is directed coaxially relative to the magnetic field of the plasma source, and is coaxial with respect to the plasma jet. An optical visualization of the plasma generated under first orientation is presented in Figure 1.2b. It may be seen that the arc plasma follow the magnetic field lines near the poles of the substrate magnetic field; in addition, a torus-shaped plasma is produced as a result of background gas ionization in the crossed field system (here the substrate was also negatively biased relative to the walls of the reactor, with the latter being grounded). An optical image of the plasma for the second orientation is depicted in Figure 1.2c; here, the plasma is focused to the magnetic pole of the substrate, and a very high current density is provided in this area.

Electrostatic confinement is also widely applied in plasma technology and electric propulsion [14, 15] as an alternative for the magnetic confinement. The multiple reflection of the charged particles from the walls charged with the opposite sign is the essence of the approach; *hollow cathode* discharge is the most developed embodiment of the concept, with a thin-walled long cylinder as the most used shape of the hollow cathode [16, 17]. Here, significantly larger plasma densities may be obtained compared with the conventional glow discharge. This fact is explained by the mechanism of the electron motion: the electrons gain the energy in the electrostatic sheath developed near the walls of the cathode, and undergo a long path to the exit from the cathode through the multiple reflections from the charged walls. Here, the probability of the ionizing collision is very high even at the low gas pressure. The secondary electrons generated from the cathode as a result of the ion bombardment of its surface, are accelerated in the sheath and become "hot" electrons; these hot electrons oscillate between the charged sheathes, and in this process produce lower temperature plasma electrons [4, 18]. Figure 1.3 shows a system with a segmented hollow cathode discharge proposed by Gallo et al., for surface engineering [17].

To provide the hollow cathode operation, the following condition should be met [9, 19]:

$$L < \lambda_{er} = \lambda_{el} \left[\frac{2m}{M} + \frac{v_{ee}}{v_m} + \frac{2}{3} \left(\frac{e\varepsilon_{ex}}{kT_e} \right) \frac{v_{ex}}{v_m} + \frac{2}{3} \left(\frac{e\varepsilon_{iz}}{kT_e} \right) \frac{v_{iz}}{v_m} + 3 \frac{v_{iz}}{v_m} \right]^{-1/2}, \qquad (1.6)$$

where L is the discharge gap between the walls; λ_{er} is the electron energy relaxation length; λ_{el} is the mean free path that electron travels prior to an elastic collision with an atom; v_{ee}/v_m is the ratio that describes the loss of energy by an electron on electron—electron coulomb collisions; $2m/M$ is proportional to the energy of the electron lost on elastic collisions with the background gas; the third and fourth terms in the equation describe the energy lost by the electron as a result of ionization and excitation, respectively; and the last term reflects the energy lost at the boundaries; $v_{ee}, v_m, v_{ex}, v_{iz}$ represent the respective frequencies of electron-electron and electron-atom elastic collisions, excitation, and ionization.

The following relation between the density n_{e0} of the plasma electrons (which approximately equals to the density of plasma in the discharge center) and density n_{h0} of the hot

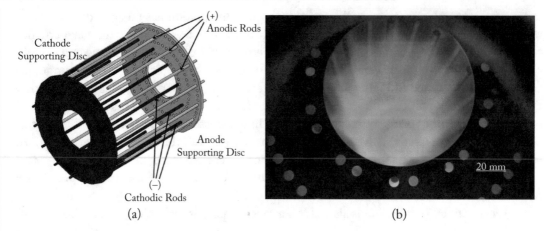

Figure 1.3: Electrostatic confinement: a schematic of the segmented hollow cathode experimental setup (a) and a photo of an electrostatic plasma confinement (b). Reprinted with permission from Gallo et al., *IEEE Trans. Plasma Sci.*, 39:3028, 2011. Copyright 2011 IEEE.

electrons can be deduced [18]:

$$n_{e0} = \frac{k_{iz}n_a}{8D_a}L^2 n_{h0},$$ (1.7)

where k_{iz} is the rate of ionization for the hot electrons; n_a is the background gas density; and D_a is the coefficient that describes the ambipolar diffusion.

The plasma density n_{e0} is much greater than the density n_{h0}. The former is inversely proportional to the diffusion coefficient D_a that can additionally be influenced by the *transverse magnetic field* to confine the electrons.

A photograph of the magnetically enhanced hollow cathode discharge is shown in Figure 1.4.

Here, the presence of the magnetic field greatly affects the geometrical structure of the plasma, thus allowing for the treatment of large surface areas with the flux of gas ions. Under these conditions, a coexistence of the conventional and magnetically-enhanced glow discharges may be seen. The former is conditioned by the emission of the electrons from the sharp edges of the hollow cathode structure, while the latter depends on the parameters of the magnetic trap with the arched configuration.

1.2   RADIO-FREQUENCY (RF) PLASMA

Instead of the direct motion of the hot (igniting) electrons toward anode, an oscillatory motion can be used to ensure efficient coupling of the electrons with the background gas, and thus to sustain the *radio-frequency (RF) plasmas* for the technological purposes [20]. Here, before reaching the electrode, electrons oscillate and undergo numerous ionization collisions with

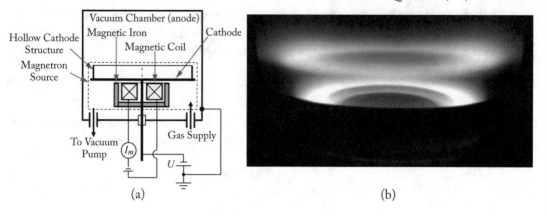

(a) (b)

Figure 1.4: Magnetically enhanced hollow cathode discharge: a schematic of the setup (a) and a photograph of the discharge (b). The bright torus inside the hollow cathode structure is conditioned by the magnetron discharge, while the dim torus above the structure is generated due to the emission of the secondary electrons from the sharp edges of the hollow cathode structure in a conventional DC glow discharge. Photo by the authors.

the background gas. At the typical RF frequencies of 1–200 MHz, low-energy ions as well as electrons with the energies over a wide range could be affected [21]. Unlike the DC plasmas, where the ion energy is controlled by applying a negative bias to a selected electrode because the floating potential developed at the electrically isolated electrode cannot generate a sufficiently strong electric field, the RF sheath modulation allows generating a strong negative floating potential [9, 22, 23].

For the simplest model of RF sheaths that does not take into account any of the ion dynamics, the electric field of an electromagnetic wave with a small amplitude that propagates in the z-direction is [9]

$$E_x = Re\left[\tilde{E}_x \exp\left\{i\omega\left(t - n_{ref}z/c\right)\right\}\right], \tag{1.8}$$

where c is the speed of light; ω is the wave frequency; and n_{ref} is the complex refractive index. The latter can be defined as $n_{ref} = n_r + in_{im}$.

The plasma electron frequency $\omega_{pe} = (ne^2/m\epsilon_0)^{1/2}$ as well as the frequency ν_m of the electron collisions with the gas neutrals are the parameters which determine the propagation of the wave; it is typical for the electron frequency to lie in the GHz range.

When $\omega > \omega_{pe}$ at the gas pressures below 100 Pa, the frequency of collision is typically small with respect to the wave frequency, $\nu_m/\omega << 1$, and $n_{im} \approx 0$ so

$$n_r \approx \left(1 - \frac{\omega_{pe}^2}{\omega^2}\right)^{1/2}. \tag{1.9}$$

In this case, the wave propagates in the plasma with very weak attenuation.

When $\omega < \omega_{pe}$ and $v_m/\omega << 1$, we obtain $n_{im} \approx \frac{\omega_{pe}}{\omega}$ and $n_r \approx 0$. For this case, the wave decays with a characteristic scale called the inertial skin depth:

$$\delta \approx \frac{c}{\omega n_{im}} = \frac{c}{\omega_{pe}}. \tag{1.10}$$

When the pressure is rather high and the condition $v_m/\omega >> 1$ is satisfied, the collisions result in high resistivity, and the resistive skin depth and refractive index are

$$\delta \approx \frac{c}{\omega n_{im}} = \frac{c}{\omega_{pe}} \left(\frac{2v_m}{\omega} \right)^{1/2}, \tag{1.11}$$

$$n_{im} \approx \frac{\omega_{pe}}{\omega} \left(\frac{\omega}{2v_m\omega} \right)^{1/2}. \tag{1.12}$$

The coupling between the power of the RF generator and the electrons can be differentiated into three modes, namely the electrostatic (E), the evanescent electromagnetic (H), and the propagating wave (W) modes.

1.2.1   CAPACITIVELY COUPLED PLASMAS (CCPS)

CCPs usually operate at 13.56 MHz in the E-mode. The reactor typically consists of two parallel electrodes fixed at several cm from each other, and biased by an RF power supply [4, 24]. An increase in the applied RF voltage V_0 results in a linear increase in the density of plasma, with the latter exhibiting a quadratic dependence on the applied frequency ω [9]:

$$n_0 = \left[\frac{\varepsilon_0 K_{stoc} K_{cap} (mM)^{1/2}}{4eh_l \varepsilon_T (T_e)} \right] \omega^2 V_0, \tag{1.13}$$

where K_{stoc} and K_{cap} represent constants that take values of $K_{cap} = 0.613$, $K_{stoc} = 0.72$ and of $K_{cap} = 0.715$, $K_{stoc} = 0.8$ for for collisionless and collisional sheaths, respectively, and $\epsilon_T (T_e)$ is a sum that can be expressed as

$$\varepsilon_T (T_e) = \varepsilon_{iz} + \frac{K_{ex}}{K_{iz}} \varepsilon_{ex} + \frac{3m}{M} \frac{K_{el}}{K_{iz}} kT_e + 2kT_e + eV_w, \tag{1.14}$$

where k is Boltzmann constant, ϵ_{iz} and ϵ_{ex} are atom ionization and excitation energies, respectively, and V_w is a potential drop at the reactor walls.

Unfortunately, it is difficult to control the density of plasma and the sheath voltage independently when using a single-frequency CCP; these parameters are primarily determined by the RF power [25, 26]. However, the RF plasma discharges with dual frequency allow one to overcome this limitation: here, one frequency is applied for plasma generation, while another is responsible for controlling the ion energy [27].

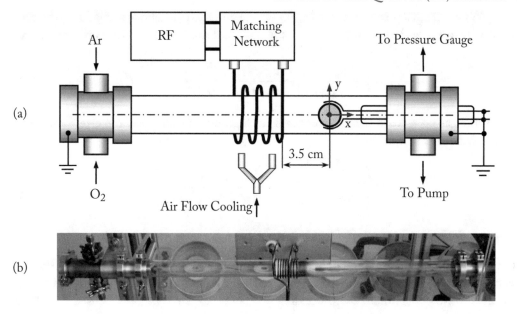

Figure 1.5: Typical setup for radio-frequency (13.56 MHz) powered nanosynthesis in the oxygen plasma: the schematic (a) and photograph of the discharge (b). Reprinted with permission from Filipič et al., *Phys. Plasmas*, 21:113506, 2014. Copyright 2014 AIP.

In CCP, the magnetic field also can play an important role: when directed parallel to the electrodes, it enhances greatly the effectiveness of so-called magnetically enhanced reactive ion etchers (MERIE) [28, 29].

1.2.2 INDUCTIVE COUPLED PLASMAS (ICPS)

RF power can be coupled to the plasma electrons not only through the electrode immersed into the reactor but through a dielectric window by use of RF coil placed outside the reactor; in this way the ***inductive coupled plasmas*** (ICPs) are operated both in E- and in H-modes. At the same time, a substrate with the treated material can be powered independently of the plasma generating loop, opening up the possibility of separately controlling the energy and flux of the ions in the ICP reactors [9, 30].

In one of the simplest design, shown in Figure 1.5 [31], the ICP discharge is generated in a cylindrical quartz tube, and RF sinusoidal current is applied:

$$I_{RF}(t) = Re\left[\tilde{I}_{RF}\exp\{i\omega t\}\right]. \tag{1.15}$$

When the gas pressure is low ($v_m/\omega << 1$) and the plasma density is high, the electron density increases with the number of turns N and does not depend on the wave frequency ω:

$$n_e = \left[\frac{\pi r_0 N^2 v_m \, (m/\varepsilon_0)^{1/2}}{4 u_B \left(h_l \pi r_0^2 + h_{r0} \pi r_0 l \right) e \varepsilon_T \left(T_e \right) lc} \right]^{2/3} I_{coil}^{4/3}. \tag{1.16}$$

When the gas pressure is high ($v_m/\omega >> 1$) at high plasma densities, the electron density increases with the frequency ω:

$$n_e = \left[\frac{\pi r_0 N^2 \, (2\omega v_m)^{1/2} \, (m/\varepsilon_0)^{1/2}}{4 u_B \left(h_l \pi r_0^2 + h_{r0} \pi r_0 l \right) e \varepsilon_T \left(T_e \right) lc} \right]^{2/3} I_{coil}^{4/3}. \tag{1.17}$$

1.3   MICROWAVE (MW) PLASMA

The microwave plasma reactors operate at a typical microwave frequency of 2.45 GHz. These reactors share a similar structure to that used in the ICP plasmas: a waveguide is used to connect the microwave generator to the processing chamber. A quartz window is used to deliver the microwave energy to the chamber to ignite and sustain the plasma discharge.

Use of electromagnetic waves in the frequency range from 300 MHz to 300 GHz to ignite plasma has been known for years [32]. In this kind of plasma, a dependence of the energy ϵ_t transferred to a species with a charge q and mass m in an oscillating electrical field characterized by the frequency ω can be defined as [33]

$$\varepsilon_t \sim \frac{q}{m} \frac{\omega}{\omega^2 + v_c^2}, \tag{1.18}$$

where v_c is a collision frequency, the value of which is determined by the pressure of the background gas.

A wide range of processes employ non-equilibrium low-pressure plasmas generated by the microwave discharges at pressures from approximately 10^{-2} to 30 kPa [34, 35].

When applied along the length of the cylinder-shaped MW reactor, the magnetic field greatly changes the processes within it, and **helicon plasmas** [36] are generated in the W-mode. The helicon wave typically propagates with a frequency of 13.56 MHz when generated in argon with $n_e = 10^{18} m^{-3}$ and $B_0 \approx 10$ mT. The magnetic field enables the propagation of electromagnetic waves when $\omega << \omega_{pe}$ [37–39]. The electron ω_{ce} and ion ω_{ci} cyclotron frequencies determine the dispersion relations for the wave propagation:

(1) when $\omega_{ce} << \omega_{pe}$, the wave with right-hand polarization (RHP) can propagate along the magnetic field lines:

$$n_{ref} = \frac{\omega_{pe}}{(\omega \omega_{ce})^{1/2} \left(1 + \frac{\omega_{ci}}{\omega} - \frac{\omega}{\omega_{ce}} \right)^{1/2}}; \tag{1.19}$$

(2) when $\omega < \omega_{ce}$, the dispersion relation with $\omega_{ci} << \omega$ can be expressed as

$$n_{ref} = \frac{\omega_{pe}}{(\omega\omega_{ce})^{1/2} \left(1 - \frac{\omega}{\omega_{ce}}\right)^{1/2}}; \quad \text{and} \tag{1.20}$$

(3) when $\omega_{ci} << \omega < \omega_{ce}$, the expression takes form of the dispersion relation for the helicon waves

$$n_{ref} = \frac{\omega_{pe}}{(\omega\omega_{ce})^{1/2}}. \tag{1.21}$$

Under conditions where the frequency of the wave approximates that of the electron cyclotron frequency $\omega = \omega_{ce}$, the resonant heating is effective and the rotation of the electrons is synchronous to the electric field wave. The process is employed in the *electron cyclotron* resonance (ECR) plasma reactors, where the magnetic field of 87.5 mT is applied to fit the frequency of 2.45 GHz. In the experiment conducted by Weatherford et al. [40], the microwaves are introduced in a section of the waveguide, with two vertically oriented SmCo magnets used to establish an ECR heating zone, as shown in Figure 1.6a. The discharge is ignited in He at 100 W of MW power and the pressure of 130 mtorr. The application of a positive bias to the electrode results in the extraction of the electrons from the ECR plasma through the aperture.

1.4   ARC PLASMAS

All of the above discussed discharges require the presence of background gas in the plasma reactor, and the gas ions are produced as a result of the ionization process. To get the metal ions, one should first evaporate the metal, and just after that the DC, RF, or MW discharge mechanisms may be launched. However, there are certain types of discharges that allow one to directly obtain the plasmas from solids, namely, the **arc discharge**, which makes it extremely suitable for the plasma-enabled material processing involving metals. The initiation of the arc plasma occurs in a region called a *cathode spot* that comprises an area of a cathode and the area where dense plasma is generated [41]; optical measurements resolve the density of the spot plasma as high as 10^{20} cm^{-3} in the arc stage of discharge [42]. The cathode spot initiation is supported by the evaporation of the cathode material, vapor ionization, and electron emission [43]. For copper cathode, the rate of the metal vapor generation can be as high as 0.05–0.1 mg/C [44].

The material is emitted not only in the form of the vapors, but as liquid droplets or solid clusters, and the velocity of the droplets can be as high as 600 m/s [45, 46]. When impinging on the surface being processed, the microparticles may become attached thus serving as precursors for the future nanogrowth.

In arc discharges, plasma interacts with the cathode in the spot region [44]. Unlike the glow discharge, the heating but not sputtering is the primary process for the emission of the cathode material. As a result of the heating and effects of the electric field, a flux of the primary electrons is emitted and accelerated in the sheath. These electrons cause heating and ionization of

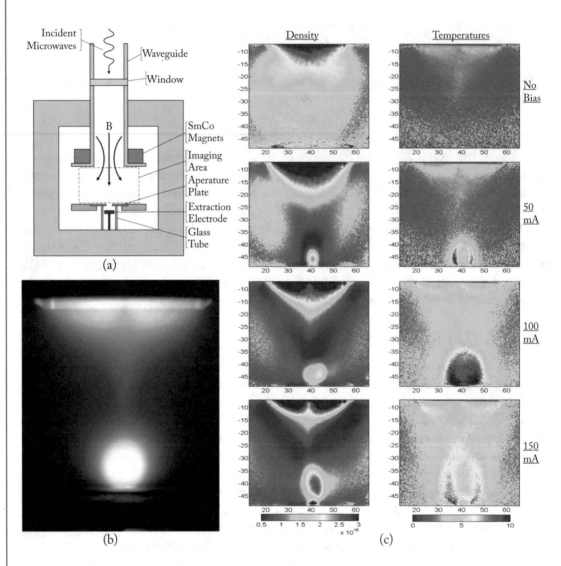

Figure 1.6: **ECR plasma cathode setup:** a schematic (a); a photograph of the discharge during operation at 100 mA (b); and the calculated electron density (cm^{-3}) and electron temperature (eV) at the current of 0, 50, 100, and 150 mA (c). Numbers along the borders of the image indicate position (mm). Reprinted with permission from Weatherford et al., *IEEE Trans. Plasma Sci.*, 39(11):2486–2487, 2011. Copyright 2011 IEEE.

the evaporated material, and again two groups of electrons are generated: high energy electrons from the cathode and slower electrons generated in the plasma. The cathode heating depends on the ion flux and the electron backflux, while the difference between the flows of evaporated atoms from the cathode and the heavy particles from the plasma determines the flow of erosion material [47]. Mechanisms of thermionic emission enhanced by the Schottky effect [44] are the foundation for the arc discharge; the temperature T_0 at the cathode surface and electric field E are the main factors determining the fluxes of the evaporated atoms $\varphi_{n0}(T)$ and emitted electrons $\varphi_{e0}(T, E)$:

$$
Y = \frac{\varphi_{e0}(T, E)}{\varphi_{n0}(T)} = AT_0^{5/2} \exp \left[\frac{eW_{evap}(1 - \chi)}{k_B T_0} \right],
\tag{1.22}
$$

where A is a constant that is dependent on the material; $\chi = W/W_{evap}$; W_{evap} is the energy of evaporation (in eV); and W is the Schottky reduced work function.

A ratio between the components of the arc voltage drop $U_a = U_p + U_c$, where U_p is the plasma potential, and U_c is the potential drop across the sheath, strongly depends on the time moment of the discharge evolution [44]. Thus, the initial cathode potential drop is about 100 V for the time duration $\tau = 14$ ns, and then it decreases to a stationary mode of about 12 V [48]. Here, the cathode potential drop exhibits a weak dependence on the arc currents, if the latter exceeds 100 A. The initial plasma is fully ionized during 30–40 ns, and then the ion fraction decreases to $\alpha_i \sim 0.5$–0.6 due to the interaction with the evaporated material [49]. This fact is used in vacuum arc plasma thrusters, where the short pulses are more effective than stationary mode [50, 51].

A diagram of a typical vacuum arc plasma source and the photographs of the discharge in operation are shown in Figure 1.7. This is an example of the arc source with the magnetic stabilization of the arc, when two magnetic coils are powered to retain the arc spot on the front surface of the evaporated cathode (the focusing coil), and to guide the arc plasma from the source to the treated parts located on the substrate (the guiding coil). Both coils are powered at the moment of the discharge generation, when a spark is ignited by use of the igniting electrode.

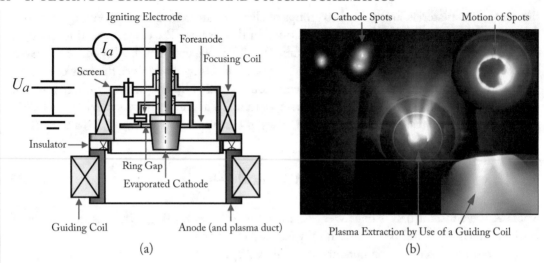

Figure 1.7: Vacuum arc discharge: a schematic of a vacuum arc plasma source with magnetic stabilization of a cathode spot (a); and photographs of the plasma source operation (b). Reprinted with permission from Baranov et al., *Rev. Mod. Plasma Phys.*, 3:7, 2019. Copyright 2019 Springer.

1.5   REFERENCES

[1] Braithwaite, N. St. J. 2000. Introduction to gas discharges, *Plasma Sour. Sci. Technol.*, 9:517. DOI: 10.1088/0963-0252/9/4/307. 5

[2] Dyatko, N. A., Ionikh, Y. Z., Meshchanov, A. V., and Napartovich, A. P. 2011. Steady-state partially constricted glow discharge, *IEEE Trans. Plasma Sci.*, 39:2532–2533. DOI: 10.1109/tps.2011.2136348. 5

[3] Zolfaghari, P., Khaledian, H. R., Aliasgharlou, N., Khorram, S., Karimi, A., and Khataee, A. 2019. Facile surface modification of immobilized rutile nanoparticles by non-thermal glow discharge plasma: Effect of treatment gases on photocatalytic process, *Appl. Surf. Sci.*, 490:266–277. DOI: 10.1016/j.apsusc.2019.06.077. 5

[4] Lieberman, M. A. and Lichtenberg, A. J. 2005. *Principles of Plasma Discharges for Materials Processing*, New York, Wiley. DOI: 10.1002/0471724254. 5, 9, 12

[5] Helander, P. 2014. Theory of plasma confinement in non-axisymmetric magnetic fields, *Rep. Prog. Phys.*, 77:087001. DOI: 10.1088/0034-4885/77/8/087001. 5

[6] Boeuf, J.-P. 2017. Tutorial: Physics and modeling of Hall thrusters, *J. Appl. Phys.*, 121:011101. DOI: 10.1063/1.4972269. 5

[7] Knapp, D. R. 2015. Planar geometry inertial electrostatic confinement fusion device, *J. Phys., Conf. Ser.*, 591:012018. DOI: 10.1088/1742-6596/591/1/012018. 5

[8] Kolobov, V. I. and Metel, A. S. 2015. Glow discharges with electrostatic confinement of fast electrons, *J. Phys. D, Appl. Phys.*, 48:233001. DOI: 10.1088/0022-3727/48/23/233001. 5

[9] Chabert, P. and Braithwaite, N. St. J. 2011. *Physics of Radio-Frequency Plasmas*, New York, Cambridge. DOI: 10.1017/cbo9780511974342. 6, 9, 11, 12, 13

[10] Chen, F. 1984. *Introduction to Plasma Physics and Controlled Fusion*, New York, Plenum. DOI: 10.1007/978-1-4757-5595-4. 7

[11] Morozov, A. I. and Savelyev, V. V. 2000. Fundamentals of stationary plasma thruster theory. Kadomtsev, B. B. and Shafranov, V. D. (Eds.), *Rev. Plasma Physics*, vol. 21, Springer, Boston, MA. DOI: 10.1007/978-1-4615-4309-1. 7

[12] Baranov, O. and Romanov, M. 2008. Current distribution on the substrate in a vacuum arc deposition setup plasma process, *Plasma Process. Polym.*, 5:256. DOI: 10.1002/ppap.200700160. 7

[13] Levchenko, I., Romanov, M., and Korobov, M. 2004. Plasma jet interaction with a spherical target in magnetic field, *IEEE Trans. Plasma Sci.*, 32(5):2139–2143. DOI: 10.1109/tps.2004.835527. 8

[14] Lim, J. W. M., Levchenko, I., Huang, S., Xu, L., Sim, R. Z. W., Yee, J. S., Potrivitu, G.-C., Sun, Y., Bazaka, K., Wen, X., Gao, J., and Xu, S. 2019. Plasma parameters and discharge characteristics of lab-based krypton-propelled miniaturized Hall thruster, *Plasma Sour. Sci. Technol.*, 28:064003. DOI: 10.1088/1361-6595/ab07db. 9

[15] Baranov, O. O., Xu, S., Xu, L., Huang, S., Lim, J. W. M., Cvelbar, U., Levchenko, I., and Bazaka, K. 2018. Miniaturized plasma sources: Can technological solutions help electric micropropulsion?, *IEEE Trans. Plasma Sci.*, 46(2):230–238. DOI: 10.1109/tps.2017.2773073. 9

[16] Hagelaar, G. J. M., Mihailova, D. B., and van Dijk, J. 2010. Analytical model of a longitudinal hollow cathode discharge, *J. Phys. D, Appl. Phys.*, 43:465204. DOI: 10.1088/0022-3727/43/46/465204. 9

[17] Gallo, S. C., Crespi, A. E., Cemin, F., Figueroa, C. A., and Baumvol, I. J. R. 2011. Electrostatically confined plasma in segmented hollow cathode geometries for surface, *Eng. IEEE Trans. Plasma Sci.*, 39(11):3028–3029. DOI: 10.1109/TPS.2011.2141690. 9

[18] Baranov, O., Xu, S., Ostrikov, K., Wang, B. B., Cvelbar, U., Bazaka, K., and Levchenko, I. 2018. Towards universal plasma-enabled platform for the advanced nanofabrication: Plasma physics level approach, *Rev. Mod. Plasma Phys.*, 2:4. DOI: 10.1007/s41614-018-0016-7. 9, 10

[19] Godyak, V. A. 2006. Nonequilibrium EEDF in gas discharge plasmas, *IEEE Trans. Plasma Sci.*, 34(3):755–766. DOI: 10.1109/tps.2006.875847. 9

[20] Chen, F. F. and Chang, J. P. 2002. *Lecture Notes on Principles of Plasma Processing*, New York, Plenum/Kluwer Publishers. DOI: 10.1007/978-1-4615-0181-7. 10

[21] Perret, A., Chabert, P., Jolly, J., and Booth, J.-P. 2005. Ion energy uniformity in high-frequency capacitive discharges, *Appl. Phys. Lett.*, 86:021501. DOI: 10.1063/1.1848183. 11

[22] Kawata, H., Yasuda, M., and Hirai, Y. 2008. Sheath voltage estimation for inductively coupled plasma etcher by impedance analysis, *Jpn. J. Appl. Phys.*, 47(8):6914–6916. DOI: 10.1143/jjap.47.6914. 11

[23] Rahman, M. T. and Dewan, M. N. A. 2015. Analytical determination of collision-less sheath properties for triple frequency capacitively coupled plasma, *Plasma Sci. Tech.*, 17(2):141–146. DOI: 10.1088/1009-0630/17/2/08. 11

[24] Rauf, S., Bera, K., and Collins, K. 2010. Power dynamics in a low pressure capacitively coupled plasma discharge, *Plasma Sour. Sci. Technol.*, 19:015014. DOI: 10.1088/0963-0252/19/1/015014. 12

[25] Diomede, P. and Economou, D. J. 2014. Kinetic simulation of capacitively coupled plasmas driven by trapezoidal asymmetric voltage pulses, *J. Appl. Phys.*, 115:233302. DOI: 10.1063/1.4884017. 12

[26] Wilczek, S., Trieschmann, J., Schulze, J., Schuengel, E., Brinkmann, R. P., Derzsi, A., Korolov, I., Donko, Z., and Mussenbrock, T. 2015. The effect of the driving frequency on the confinement of beam electrons and plasma density in low-pressure capacitive discharges, *Plasma Sour. Sci. Technol.*, 24:024002. DOI: 10.1088/0963-0252/24/2/024002. 12

[27] Lieberman, M. A., Booth, J. P., Chabert, P., Rax, J. M., and Turner, M. M. 2002. Standing wave and skin effects in large-area, high-frequency capacitive discharges, *Plasma Sour. Sci. Technol.*, 11:283. DOI: 10.1088/0963-0252/11/3/310. 12

[28] Kushner, M. J. 2003. Modeling of magnetically enhanced capacitively coupled plasma sources: Ar discharges, *J. Appl. Phys.*, 94(3):1436–1447. DOI: 10.1063/1.1587887. 13

[29] Yang, S., Zhang, Y., Wang, H.-Y., Wang, S., and Jiang, W. 2017. Electrical asymmetry effects in magnetized capacitively coupled plasmas in argon, *Plasma Sour. Sci. Technol.*, 26:065011. DOI: 10.1088/1361-6595/aa6ef1. 13

[30] Godyak, V. 2013. Ferromagnetic enhanced inductive plasma sources, *J. Phys. D, Appl. Phys.*, 46:283001. DOI: 10.1088/0022-3727/46/28/283001. 13

[31] Filipič, G., Baranov, O. Mozetič, M., Ostrikov, K., and Cvelbar, U. 2014. Uniform surface growth of copper oxide nanowires in radiofrequency plasma discharge and limiting factors, *Phys. Plasmas*, 21:113506. DOI: 10.1063/1.4901813. 13

[32] Mac Donald, A. D. 1966. *Microwave Breakdown in Gases*, New York, John Wiley & Sons. 14

[33] Szabó, D. V. and Schlabach, S. 2014. Microwave plasma synthesis of materials—from physics and chemistry to nanoparticles: A materials scientist's viewpoint, *Inorganics*, 2:468–507. DOI: 10.3390/inorganics2030468. 14

[34] Lebedev Yu, A. 2015. Microwave discharges at low pressures and peculiarities of the processes in strongly non-uniform plasma, *Plasma Sour. Sci. Technol.*, 24:053001. DOI: 10.1088/0963-0252/24/5/053001. 14

[35] Lebedev Yu, A., Epstein, I. L., Tatarinov, A. V., and Shakhatov, V. A. 2010. Electrode microwave discharge: Areas of application and recent results of discharge physics, *J. Phys., Conf. Ser.*, 207:012002. DOI: 10.1088/1742-6596/207/1/012002. 14

[36] Shinohara, S. 2018. Helicon high-density plasma sources: Physics and applications, *Adv. Phys.: X*, 3(1):1420424. DOI: 10.1080/23746149.2017.1420424. 14

[37] Stenzel, R. L. and Urrutia, J. M. 2015. Helicons in unbounded plasmas, *Phys. Rev. Lett.*, 114:205005. DOI: 10.1103/physrevlett.114.205005. 14

[38] Chen, F. F. 2012. Performance of a permanent-magnet helicon source at 27 and 13 MHz, *Phys. Plasmas*, 19:093509. DOI: 10.1063/1.4754580. 14

[39] Furukawa, T., Takizawa, K., Kuwahara, D., and Shinohara, S. 2017. Study on electromagnetic plasma propulsion using rotating magnetic field acceleration scheme, *Phys. Plasmas*, 24:043505. DOI: 10.1063/1.4979677. 14

[40] Weatherford, B. R., Barnat, E. V., and Foster, J. E. 2011. Two-dimensional LCIF images of electron density and temperature within an ECR plasma cathode, *IEEE Trans. Plasma Sci.*, 39(11):2486–2487. DOI: 10.1109/tps.2011.2158120. 15

[41] Hantzsche, E. 2003. Mysteries of the arc cathode spot: A retrospective glance, *IEEE Trans. Plasma Sci.*, 31(5):779. DOI: 10.1109/tps.2003.818412. 15

[42] Methling, R., Popov, S. A., Batrakov, A. V., Uhrlandt, D., and Weltmann, K.-D. 2013. Spectroscopic investigation of a Cu—Cr vacuum arc, *IEEE Trans. Plasma Sci.*, 41(8):1904. DOI: 10.1109/tps.2015.2443856. 15

[43] Beilis, I., Koulik, Y., Yankelevich, Y., Arbilly, D., and Boxman, R. L. 2015. Thin-film deposition with refractory materials using a vacuum arc, *IEEE Trans. Plasma Sci.*, 43(8):2223–2328. DOI: 10.1109/tps.2015.2432577. 15

[44] Beilis, I. I. 2001. State of the theory of vacuum arcs, *IEEE Trans. Plasma Sci.*, 29(5):657. DOI: 10.1109/27.964451. 15, 17

[45] Wang, L., Zhang, X., Wang, Y., Yang, Z., and Jia, S. 2018. Simulation of cathode spot crater formation and development on CuCr alloy in vacuum arc, *Phys. Plasmas*, 25:043511. DOI: 10.1063/1.5023213. 15

[46] Zhang, X., Wang, L., Jia, S., and Shmelev, D. L. 2017. Modeling of cathode spot crater formation and development in vacuum arc, *J. Phys. D, Appl. Phys.*, 50:455203. DOI: 10.1088/1361-6463/aa8db3. 15

[47] Beilis, I. 2003. The vacuum arc cathode spot and plasma jet: Physical model and mathematical description, *Contrib. Plasma Phys.*, 43(3–4):224–236. DOI: 10.1002/ctpp.200310018. 17

[48] Beilis, I. I. 2010. The nature of high voltage initiation of an electrical arc in a vacuum, *Appl. Phys. Lett.*, 97:121501. DOI: 10.1063/1.3491446. 17

[49] Beilis, I. I. 2013. Cathode spot development on a bulk cathode in a vacuum arc, *IEEE Trans. Plasma Sci.*, 41(8):1979. DOI: 10.1109/TPS.2013.2256472. 17

[50] Baranov, O., Levchenko, I., Xu, S., Wang, X. G., Zhou, H. P., and Bazaka, K. 2019. Direct current arc plasma thrusters for space applications: Basic physics, design and perspectives, *Rev. Mod. Plasma Phys.*, 3:7. DOI: 10.1007/s41614-019-0023-3. 17

[51] Sun, Y., Levchenko, I., Lim, J. W. M., Xu, L., Huang, S., Zhang, Z., Thio, F., Potrivitu, G.-C., Rohaizat, M., Cherkun, O., Chan, C. S., Baranov, O., Bazaka, K., and Xu, S. Miniaturized rotating magnetic field driven plasma system: proof-of-concept experiments. *Plasma Sources Sci. Technol*, in press (2020). DOI: 10.1088/1361-6595/ab9b34 17

C H A P T E R   2

Plasma Parameters with Respect to Material Processing

ABSTRACT

This chapter defines key parameters of plasmas generally used for the synthesis and processing of materials. A general approach to modeling such systems is briefly described, with main emphasis on ion energy and ion current density as prime factors that control parameters of plasma-based processing and treatment of materials.

When considering the possible effects of plasmas onto a surface, three main processes can be distinguished: (i) *deposition*, when a layer of some material is added onto the existing surface; (ii) *modification*, when the morphology and chemical composition of the under-surface layers are changed without changing the geometrical features of the surface; and (iii) *sputtering*, when the surface layers are removed through the ion bombardment (a variation of the process where a chemical reactant is used is called *etching*). However, the real situation is often far more complicated, with all of the above processes present during the plasma treatment yet with the different intensity. A transition from one type of the surface processing to another depends on the type of particles involved in the plasma-surface interactions, the particle energy, and flux [1, 2].

Ion energy and density of ion current as factors that define the nature of the surface treatment. The plasma chemical compositions, as well as the ratio between the charged, excited, neutral particles, and radicals depend greatly on the discharge type, so it should be chosen carefully. Different types of plasma discharges allow adjusting the plasma density by tuning the adsorbed power within the limits specified by the discharge type and the equipment.

Plasma density depends on the power through the following relation [3]:

$$n_0 = \frac{P_{abs}}{e u_B S_0 \varepsilon_T} = \frac{P_{abs}}{e^{3/2} (T_e/M)^{1/2} S_0 \varepsilon_T}, \tag{2.1}$$

where u_B is a Bohm velocity; ε_T is the total energy lost per electron-ion pair that disappear from the system; and S_0 is the effective area for particle loss.

In general, with respect to the rate of change of the surface geometrical size V_g (m/s), the effect of the ion flux can be described as

$$V_g \approx \frac{j_i}{e} a_0^3 k_i \left(\varepsilon_i\right) = 0.61 n_{0u} B a_0^3 k_i \left(\varepsilon_i\right) = 0.61 \frac{P_{abs}}{e S_0 \varepsilon_T} a_0^3 k_i \left(\varepsilon_i\right), \tag{2.2}$$

where j_i is the density of the ion current; a_0^3 is the elementary volume of material added or removed under the action of one ion; $k_i(\varepsilon_i)$ is a coefficient that is named after the specific process (sputtering, evaporation, attachment, etc.) and determines (i) the number of the volumes a_0^3 removed or added after the action of the ion and (ii) the sigh ("plus" for additive processes such as deposition, and "minus" for the material removal processes such as sputtering or etching).

In spite of the positive effect caused by the ion energy [4], which is expressed in activation and functionalization of the processed surface, high value of this parameter may result in intensive sputtering of the surface thus destroying the growth process. Fortunately, the ion energy can be easily tuned by either selecting an appropriate negative bias (when high values of the energy are necessary) or increasing the total gas pressure in the reactor. Since the high ion energy intensifies the sputtering processes, low ion energy is usually required to grow the surface structures. Depending on the plasma source used in the processing chamber, two approaches are used to obtain ions with low energy, as described below.

Ion energy and ion current density for DC-based systems.   When using DC discharges and growing the structures on electrodes, the high-pressure operation modes are applied since the nature of the DC discharge ignition does not allow for the use of the low voltage drops across the discharge gap. In this case, the ion energy can be regulated by the gas pressure, which is described by the expression:

$$\varepsilon_i = U_s \exp\left(-\frac{l_s}{\lambda_{ir}}\right) = U_s \exp\left(-l_s n_g \sigma_{ir}\right) = U_s \exp\left(-\frac{l_s P_g \sigma_{ir}}{k_B T_g}\right), \tag{2.3}$$

where l_s is the thickness of the plasma sheath near the electrode; λ_{ir} is a mean free path for the ion between the charge exchange collisions; σ_{ir} is a charge exchange cross-section; n_g is a mean density of gas particles in the reactor chamber; P_g is the total gas pressure; and T_g is the gas temperature. As we can see, the increased gas pressure results in a decrease in the ion energy.

The thickness of the plasma sheath l_s can be found for two extreme cases of the bias supply:

- for pulse bias U_s, when a matrix sheath is formed, the sheath thickness is

$$l_s = \lambda_{Ds} \left(\frac{2U_s}{T_e}\right)^{1/2} = \left(\frac{\varepsilon_0 T_e}{n_0 e}\right)^{1/2} \left(\frac{2U_s}{T_e}\right)^{1/2} = \left(\frac{2\varepsilon_0 U_s}{n_0 e}\right)^{1/2}, \tag{2.4}$$

where λ_{Ds} is the Debye length, T_e is the electron temperature (eV), n_0 is the plasma density, and ϵ_0 is the vacuum permittivity. In this case the sheath thickness can be tens of Debye lengths; and

- for DC bias U_s, when Child law sheath is formed, the sheath thickness is

$$l_s = \frac{\sqrt{2}}{3} \lambda_{Ds} \left(\frac{2U_s}{T_e} \right)^{3/4} = 0.8 \left(\frac{\varepsilon_0}{n_0 e} \right)^{1/2} \frac{U_s^{3/4}}{T_e^{1/4}}. \tag{2.5}$$

Ion energy and ion current density for RF- and MW-based systems. When growing the structures in high- or low-pressure plasmas (RF, MW), a substrate is usually either under the floating potential or is grounded; here, the ion energy is determined by the discharge conditions:

$$\varepsilon_i = \varepsilon_s \exp\left(-\frac{l_s P_g \sigma_{ir}}{k_B T_g} \right), \tag{2.6}$$

where ε_i is the energy gained by the ion in the plasma sheath, and the energy ε_s does not exceed the values from a few eV to a few tens of eV.

After the calculation of the ion energy, for the specified energy range the sputter yield can be found as

$$\Upsilon = \Upsilon_0 \left(\frac{\varepsilon_i}{U_0} \right)^k, \tag{2.7}$$

where Y_0, U_0, and k are constants specified by an "ion—sputtered material" pair.

Temperature of a substrate is also a parameter of great importance. To obtain the expression for the dependence of the substrate temperature on the other parameters, we consider a balance of heat on the surface.

The heat is provided to the surface through recombination W_{rec} and ion bombardment W_i, while it is spent on substrate heating W_{heat}, thermal conductivity through the background gas mixture W_{cond}, heat radiation W_{rad}, and molecule evaporation W_{evap}:

$$W_{rec} + W_i = W_{heat} + W_{cond} + W_{rad} + W_{evap}. \tag{2.8}$$

The surface density of the power delivered through recombination is

$$W_{rec} = e K_{rec} \varphi_r \varepsilon_{rec} = e \left(\frac{k_B T_g}{2\pi M_r} \right)^{1/2} n_r \varepsilon_{rec}, \tag{2.9}$$

where φ_r is a flux of the radicals to the surface, M_r is the radical mass, ε_r is the energy released as a result of recombination, and K_r is a numerical coefficient related to the recombination probability [5].

The power delivered through the ion bombardment is

$$W_i = j_i \varepsilon_i = 0.61 e^{3/2} \left(\frac{T_e}{M_i} \right)^{1/2} n_0 \varepsilon_i, \tag{2.10}$$

where j_i is the density of the current extracted from the plasma to the substrate, A/m^2.

The power spent on substrate heating is

$$W_{heat} = c_s \rho_s h_s \frac{dT_s}{dt},$$ (2.11)

where c_s is specific heat of the sample material, J/(kg · K); h_s is the thickness of the sample; and ρ_s is the sample material density. The power abstracted due to the thermal conductivity through the background gas is

$$W_{cond} = k_s \frac{T_s - T_0}{l_t},$$ (2.12)

where k_s is thermal conductivity of the background gas, W/(m · K); l_t is the distance between the sample and the vacuum chamber walls; and T_0 is the temperature of the reactor walls.

The power removed through the heat radiation is described by the Stefan–Boltzmann law:

$$W_{rad} = \varepsilon \sigma \left(T_s^4 - T_0^4 \right),$$ (2.13)

where ε is the emissivity of a given surface and σ is the Stefan–Boltzmann constant.

The power spent on molecule evaporation is

$$W_{evap} = e \varepsilon_{ev} \frac{\alpha P_0}{\sqrt{2\pi M k_B T_s}},$$ (2.14)

where ε_{ev} is the energy of vaporization, eV; α is a sticking coefficient; P_0 is the gas pressure, Pa; M is the mass of a particle, Kg; and T_s is the temperature, K.

2.1   REFERENCES

[1] Baranov, O., Xu, S., Ostrikov, K., Wang, B. B., Cvelbar, U., Bazaka, K., and Levchenko, I. 2018. Towards universal plasma-enabled platform for the advanced nanofabrication: Plasma physics level approach, *Rev. Mod. Plasma Phys.*, 2:4. DOI: 10.1007/s41614-018-0016-7. 23

[2] Seo, D. H., Rider, A. E., Arulsamy, A. D., Levchenko, I., and Ostrikov, K. 2010. Increased size selectivity of Si quantum dots on SiC at low substrate temperatures: An ion-assisted self-organization approach. *J. Appl. Phys.*, 107:024313. DOI: 10.1063/1.3284941. 23

[3] Lieberman, M. A. and Lichtenberg, A. J. 2005. *Principles of Plasma Discharges for Materials Processing*, New York, Wiley. DOI: 10.1002/0471724254. 23

[4] Luo, K., Zhang, Q., Yuan, H., Liu, Y., Wang, X., Zhang, J., Hu, W., Xu, M., Xu, S., Levchenko, I., and Bazaka, K. 2020. Facile synthesis of Ag/Zn$_{1-x}$Cu$_x$O nanoparticle compound photocatalyst for high-efficiency photocatalytic degradation: Insights into the synergies and antagonisms between Cu and Ag, *Ceram. Int.*, 46, (in press). DOI: 10.1016/j.ceramint.2020.06.102. 24

[5] Filipič, G., Baranov, O., Mozetič, M., Ostrikov, K., and Cvelbar, U. 2014. Uniform surface growth of copper oxide nanowires in radiofrequency plasma discharge and limiting factors, *Phys. Plasmas*, 21:113506. DOI: 10.1063/1.4901813. 25

CHAPTER 3

How to Control Plasma Parameters

ABSTRACT

This chapter describes several important technological approaches and schematic solutions for efficient control of plasma parameters in the technological setups. Incorporating and external-substrate technological architectures are introduced and discussed, including magnetically enhanced RF discharges, arrays of ferromagnetic enhanced inductive plasma sources, arrays of helicon plasma sources, and other important systems.

The large diversity of plasma-generating circuits can be divided into two main architectures: (i) a substrate acts as an electrode in the plasma-generating circuit, or (ii) a substrate is exposed to the plasma generated by an external plasma source and plays no role in the plasma-generating circuit [1]. These general structures are termed the incorporated substrate scheme (ISS) and the external substrate scheme (ESS), respectively (Figure 3.1).

The necessity to control the density of ion current and energy of ions has led to the development of various technological solutions that rely on either electrostatic or electromagnetic methods of control.

Static electric fields are widely applied to control the parameters of the ion flux, and ion extraction techniques are used to control the movement of ions. The extraction can be accomplished using a grid system such as the one used in ion implantation, or simply by applying a negative bias to a substrate, such as in plasma immersion ion implantation and deposition (PIII&D). The first method allows for collimation of the extracted ion flux and precise adjustment of their energy to the necessary energy level; yet, this comes at a cost of introducing expensive separating and steering systems. The second method is cheaper yet the ions have the wider spread in terms of their energy distribution.

Employment of the magnetic fields of a few tens of militeslas [2, 3] allows one to trap the plasma electrons but not ions, and this approach is implemented in various plasma-based technological setups [4, 5] and devices [6, 7]. As a result of the interactions of the electrons with the magnetic field, and ions—with the electrons, the electric field arises at the spatial scale much greater than that of the electrostatic sheath. This electrical field guides the motion of the plasma ions thus shaping the ion flux. In the electromagnetic control, the incorporated substrate scheme (ISS) is often applied in conjunction with the magnetic field arched above the substrate

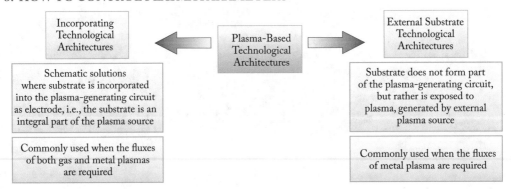

Figure 3.1: Plasma-based technological architectures with respect to the function of the substrate within the plasma-generating circuit: we distinguish the "incorporating architectures" and "external architectures." Reprinted with permission from Baranov et al., *Appl. Phys. Rev.*, 4:041302, 2017. Copyright 2017 AIP.

(Figure 1.4), while the external substrate scheme (ESS) is coupled with the guiding magnetic field like in the vacuum arc setup (Figure 1.7).

3.1   INCORPORATING TECHNOLOGICAL ARCHITECTURES

Closed drift configuration in the plasma generation region.   This configuration arises once the magnetic field of arched configuration is generated by use of a coil mounted under the negatively biased substrate, such as in the case of sputtering magnetrons (Figure 1.4). Plasma electrons are trapped in the space between the substrate (cathode) surface and the arched lines of the magnetic field. Here, the transfer of the energy gained by the elections in the sheath to the background gas is very effective, which is evidenced by a bright plasma torus corresponding to the region of highest ionization and excitation. Radius R_c of curvature of the magnetic field is a control parameter with respect to the width of the plasma torus and, correspondingly, width of the race-track appearing on the surface of the treated cathode due to the ion sputtering. The additional effect is provided by the implementation of the hollow cathode design. This is clearly illustrated with the magnetostatic FEMM^{TM} software [8]. Three possible configurations of the magnetic field over the biased cathode are shown in Figure 3.2a,d,k: with the small radius R_c and a planar cathode, a small radius and hollow cathode, and a large radius with hollow cathode, respectively [9]. Images of the plasma torus shown in Figures 3.2b,e,l demonstrate the possibility of control over plasma distribution for these configurations. At the same time, the influence of the above configurations on the distribution of the density of ion current is confirmed by the probe measurements; the results are shown in Figures 3.2c,f,m. The magnetic field strength also plays a key role in defining the discharge characteristics, and it is shaped by the iron core.

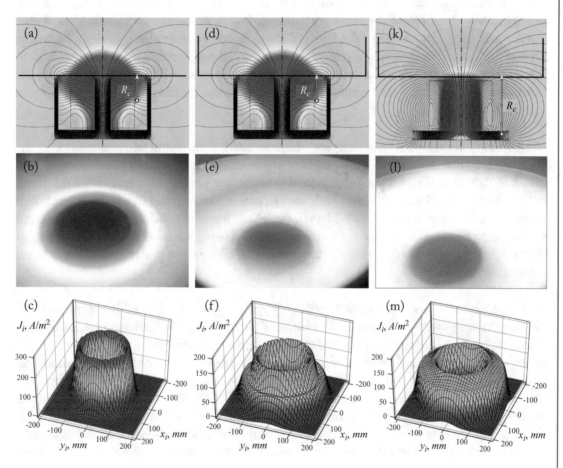

Figure 3.2: Control of plasma discharge by use of arched magnetic field in the incorporated substrate scheme (ISS): small radius R_c of curvature the magnetic field, and planar cathode (a,b,c); small radius R_c, and hollow cathode structure (d,e,f); large radius R_c, and hollow cathode structure (k,l,m). The change in the plasma discharge and distribution of the ion current is clearly visible: from the bright narrow beam (b,c) to the wide distribution (l,m). Reprinted with permission from Baranov et al., *Mater. Chem. Phys.*, 188:143, 2017. Copyright 2017 Elsevier.

For the typical configuration of the planar magnetron discharge depicted in Figures 3.2a,b,c, the plasma discharge is concentrated in the narrow racetrack under the region of very dense plasma [10–12]. Here, the location of the peak of the ion current density distribution is between the inner and outer poles of the magnetic system. When the hollow cathode structure is added to the cathode, the shape of the discharge is changed significantly yet the most intense plasma is still concentrated within the region between the magnetic system poles, as it could be seen in Figures 3.2d,e,f. After removing a part of the iron core, the magnetic field is "de-focused," so the radius R_c of the curvature of the magnetic field lines is increased two-fold. As a result, the plasma expands over the entire surface of the cathode except for the central part, as shown in Figures 3.2k,l,m [13–15].

Magnetically enhanced RF discharges.   Implementation of magnetic confinement in the CCP discharge has led to the development of the magnetically enhanced reactive ion etchers (MERIE) where a set of coils mounted around the substrate generates the magnetic field. Here, the necessity to sustain the relatively high gas pressure is omitted, and independent control of ion energy and current density is possible; yet, the lack of treatment uniformity is often observed in these setups [16]. ICP discharges can also be guided using a similar approach as shown in Figure 3.3 [17], yet it should be mentioned that in this case the setup is related to the external substrate scheme (ESS).

An RF control of the ion flux uniformity in CCP discharges.   The lack of uniformity in CCP discharges can be overcome by use of RF control of the ion flux [18]. The superposition of electromagnetic waves propagating from the substrate center to its periphery with counter propagating waves result in the formation of standing waves. The maximum of the power adsorption is observed in the center, and the effect is more prominent for large substrates at high frequencies. The electrostatic edge effects result in the maximum of the RF current at the edges of the discharge due to the small physical distance between the vacuum chamber walls and the RF electrode. The field non-uniformities due to the skin effect are prominent at high frequencies due to the increased plasma density, when the skin depth is close to the electrode gap. Tuning the applied RF power and the driving frequency affected the distribution of the ion current density in the double-frequency CCP discharges when these were generated over a large area substrates [19]. An increase in the driving frequency from 13.56–81.36 MHz changes the ion flux from one that is fairly uniformly distributed to one with a dome-shaped distribution, while increasing the RF power from 50–265 W results in the enhancement of the ion current along the outer boundary of the discharge. For low RF power, the maximum of the ion flux is located around the center of the substrate due to the effect of standing waves. Increasing the RF power results in the densification of the plasma along the sides of the discharge, with the phenomenon arising from a combination of the skin and edge electromagnetic effects.

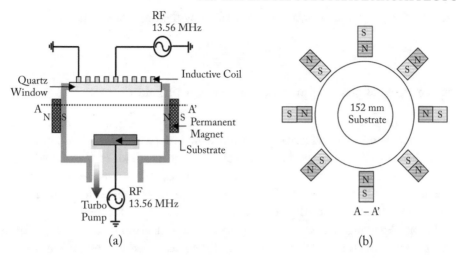

Figure 3.3: Schematic diagram of the magnetically enhanced inductively coupled plasma etch system. Set of magnetic coils mounted around the chamber (the "magnetic bucket") insulates the plasma from the walls thus enhancing the adsorbed power; the combination of capacitively coupled plasma (CCP) and inductively coupled plasma (ICP) discharges brings the advantages of lower sheath voltages, higher densities, and independent control of both of these parameters: side view (a) and top view (b) of the setup. Reprinted with permission from Kim et al., *IEEE Trans. Plasma Sci.*, 32:1362, 2004. Copyright 2004 IEEE.

3.2   EXTERNAL SUBSTRATE ARCHITECTURES

For reactors that do not integrate the substrate as part of their plasma generating circuit, and instead expose the substrate to plasma generated by an external plasma source, the magnetic field is also an effective tool to control the plasma distribution along the treated surface, yet the configuration of the magnetic traps is different. Here, the control of the ion energy is achieved by use of the PIII&D technique.

An array of ferromagnetic enhanced inductive plasma sources.   To overcome the limitation implied by low gas pressure with respect to the dense plasma generation in CCP discharge operated at the frequency of 13.56 MHz, the ICP source can be enhanced with ferromagnetic cores (FMICP), which results in generation of high-density plasma with uniform distribution of the ion flux over the large-area substrate [20, 21].

An array of helicon plasma sources.   To shape the ion flux for the purpose of controlling the density and uniformity of plasmas over large-area substrates, a set of helical sources can be used [22]. Helicon waves are related to the bounded whistler electromagnetic waves with the right-hand circular polarization, which are described by the dispersion relation [23]

$$\beta = \frac{\omega}{k} \frac{n_0 e \mu_0}{B_0},$$

(3.1)

where the density n_0 of plasma is proportional to magnetic field B_0.

A neutral loop configuration in the plasma generation region.   In this method, three electromagnetic coils are wound around a reactor within which material processing takes place. Here, the top and bottom coils are powered to generate the magnetic fields of the same direction thus creating the magnetic bottle configuration, while the middle coil is powered opposite, thus generating the neutral magnetic loop [24, 25]. When RF power is applied to an antenna coil that is mounted to be concentric with the neutral loop, the electric field is generated azimuthally to the neutral loop. Applying the RF power results in the ionization of the background gas and formation of the plasma torus with the radius of the torus controlled by tuning the current in the middle coil.

A set of magnetic mirrors along the chamber walls.   Multipole confinement is one of the oldest schemes used for plasma confinement, and is still widely employed to control the ion flux from the external plasma source. In these setups (Figure 3.3), a combination of magnetic mirrors (called "the magnetic bucket") along the vacuum chamber walls is used to isolate the plasma discharge from the walls, thus reducing the plasma loss and increasing the plasma density according to the above equation (decrease of the effective area S_0 for particle loss) [26]. Here, the use of Helmholtz type electromagnetic coils in conjunction with permanent magnets can be applied to control the parameters of the ICP discharge [27].

A multi-slot antenna and an external magnetic field to control microwave plasma.   Effective control over the microwave plasma discharge with respect to the radial distribution of the ion flux was demonstrated by Yasaka et al., by use of a multi-slotted planar antenna [28–30], where the antenna provides for uniform radiation, and the eigenmode structures of surface waves do not affect the plasma. The change in the vertical position of the antenna is used to uniformly treat the substrate as large as 700 cm^2. The diverging magnetic field is also applied to guide the electron cyclotron resonance (ECR) plasma, where the increase in the magnetic field increases the width of the spatial distribution of plasma density [31].

An open drift configuration: a magnetic mirror in the plasma generation region.   When a magnetic coil is wound around the plasma source, a magnetic mirror formed in the region where plasma is generated facilitates its expansion in the direction opposite to the magnetic field gradient [32]. This configuration is widely applied in magnetic filters used in vacuum arc plasma setups to separate plasma from the micro-droplets formed in the arc spots. Here, the use of the appropriate configuration of the wound coil within the volume of the processing chamber allows one to direct plasma to a selected area of the substrate [33].

The flexibility of plasma control using the open drift configuration of the magnetic field is enhanced by the large number of possible combinations of magnetic coils and their powering. The system that contains two magnetic mirrors with perpendicular axes was described previously [34]; see Figure 1.2. In this setup the coil is mounted under the substrate surface so the magnetic field generated by it is transverse with the magnetic field generated by a coil near the plasma source. A similar setup was developed for a cylindrical substrate, when a coil is moved along the axis of the substrate; this configuration allows for scanning the substrate surface with the ion fluxes extracted from the arc source [35].

In a coaxial configuration the coil under the substrate is positioned coaxially with respect to the coil of the plasma source, and two possible configurations of the resulting magnetic field are produced in the space between the coils.

(a) A magnetic bottle is formed when the magnetic fields of the coils are directed simultaneously. This configuration is used to focus the plasma onto the substrate and to control the ion-to-neutral ratio [36, 37]; however, the reduction of the electron conductivity across the magnetic field leads to the negative effect of this magnetic field onto the plasma source operation.

(b) A magnetic cusp or a circular mirror is formed when the magnetic fields of the coils are directed opposite to each other. Here, plasma tends to expand in a radial direction through the cusp, which is used in some configurations of the plasma filters. The influence of this configuration onto the plasma source operation modes is much less pronounced when compared to the magnetic bottle configuration.

With respect to the flexible control of the ion fluxes, a setup incorporating two complex magnetic mirrors with parallel axes is more effective [38]. In this approach, two magnetic coils (one is wound around the plasma source, and another is installed under the substrate in a peripheral region) are powered to generate the magnetic fields in the same direction. This arrangement allows creating the magnetic bottle and focusing the plasma from the source to the peripheral area of the substrate. To counteract the negative impact of the bottle configuration on the operation of the plasma source, a third coil is used, which is also mounted under the substrate as shown in Figure 3.4. The powering of this coil creates a local configuration of the magnetic cusp thus facilitating the electron transport to the walls of the vacuum chamber that are grounded [39]. In such a system, the plasma is injected from the arc source through the "neck" of the magnetic bottle.

The flexibility of control is provided by changing the powering of the coils at the specified time moments, and rotating the substrate. Different configurations of the resultant magnetic field shape the ion flux, and the time sequence of the fluxes form the time-averaged ion flux distribution along the substrate surface. For systems that use a vacuum arc plasma source, the topography of the calculated magnetic fields are shown in Figures 3.4a,d,k, the photographs of plasma discharges shown in Figures 3.4b,e,l, and the measured distributions of the density of the

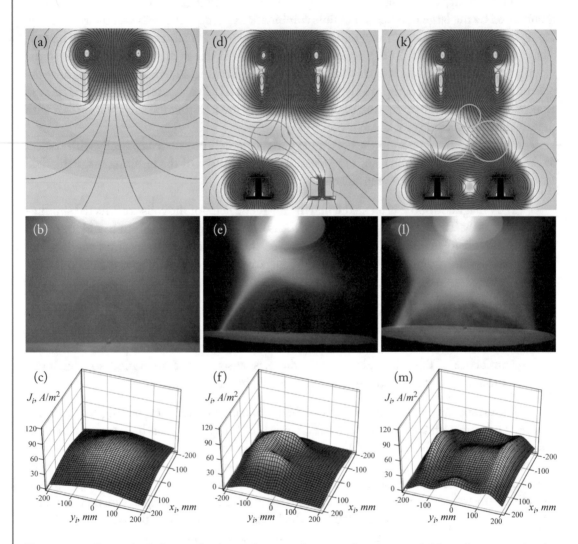

Figure 3.4: Control of plasma discharge by use of mirrored magnetic field in the external substrate scheme (ESS): a divergent magnetic field (a–c); a bottle configuration (d–f); a bottle-cusp configuration (k–m); the topography of the magnetic field (a, d, k); photographs of the arc plasma discharge (b, e, l); measured distributions of the ion current density (c, f, m). The approximation of the negative space charge regions by a set of spheres is shown by circles in (d, k). Reprinted with permission from Baranov et al., *Mater. Chem. Phys.*, 188:143, 2017. Copyright 2017 Elsevier.

ion current over substrate are shown in Figures 3.4c,f,m. It can be seen that the initial Gaussian distribution for the divergent magnetic field (Figures 3.4a–c) is changed to a horseshoe-like distribution for the bottle configuration (Figures 3.4d–f), and then to a complex five-peaked distribution for the bottle-cusped configuration (Figures 3.4k–m).

The motion of ions though a set of magnetic traps is explained within the approach of plasma optics [11]. Here, each trap is approximated by the charged spheres with the dependence of the electric potential on the radius of the trap [40]:

$$\varphi(r) = Ar^2, \tag{3.2}$$

where A is a constant.

This assumption allows describing the ion path using equations that describe the motion of a particle in a central field [41]. The simulation shows that to explain the experimental results, a strong electric field of up to 500 V/m should be produced in the region of the plasma trap due to the excess of electrons over ions of about 10^{-5} [38]. It is noteworthy that later this approach was used to explain the influence of the magnetic field on the focusing of the ion flux in the vacuum arc thruster. After the comparison of the results with the above calculation it was concluded that for the effective application of the magnetic control, the plasma density n (m^{-3}) in the magnetic trap and a spatial extent a (m) of the trap should be described by the dependence $na^3 = 10^{-14}$ [42]. To meet the condition in the vacuum arc thruster, the magnetic field should be as high as 0.1 T.

3.3   REFERENCES

[1] Baranov, O., Bazaka, K., Kersten, H., Keidar, M., Cvelbar, U., Xu, S., and Levchenko, I. 2017. Plasma under control: Advanced solutions and perspectives for plasma flux management in material treatment and nanosynthesis, *Appl. Phys. Rev.*, 4:041302. DOI: 10.1063/1.5007869. 29

[2] Wen, L., Kumar, M., Sahu, B. B., Jin, S. B., Sawangrat, C., Leksakul, K., and Han, J. G. 2015. Advantage of dual-confined plasmas over conventional and facing-target plasmas for improving transparent-conductive properties in Al doped ZnO thin films, *Surf. Coat. Technol.*, 284:85. DOI: 10.1016/j.surfcoat.2015.06.084. 29

[3] Baranov, O., Zhong, X., Fang, J., Kumar, S., Xu, S., Cvelbar, U., Mariotti, D., and Ostrikov, K. 2014. Dense plasmas in magnetic traps: Generation of focused ion beams with controlled ion-to-neutral flux ratios, *IEEE Trans. Plasma Sci.*, 42(10):2518. DOI: 10.1109/tps.2013.2295626. 29

[4] Levchenko, I., Ostrikov, K., Zheng, J., Li, X., Keidar, M., and Teo, K. 2016. Scalable graphene production: Perspectives and challenges of plasma applications, *Nanoscale*, 8:10511. DOI: 10.1039/c5nr06537b. 29

[5] Vogel, U., Klaus, C., Nobis, C., and Bartha, J. W. 2012. Analysis of the energy input during wire coating from a cylindrical magnetron source, *Thin Solid Films*, 520(20):6404. DOI: 10.1016/j.tsf.2012.05.072. 29

[6] Baranov, O., Fang, J., Keidar, M., Lu, X., Cvelbar, U., and Ostrikov, K. 2014. Effective control of the arc discharge-generated plasma jet by smartly designed magnetic fields, *IEEE Trans. Plasma Sci.*, 42(10):2464. DOI: 10.1109/tps.2014.2323263. 29

[7] De Gryse, R., Haemers, J., Leroy, W. P., and Depla, D. 2012. Thirty years of rotatable magnetrons, *Thin Solid Films*, 520(18):5833. DOI: 10.1016/j.tsf.2012.04.065. 29

[8] Bogaerts, A., Bultinck, E., Kolev, I., Schwaederl, L., Van Aeken, K., Buyle, G., and Depla, D. 2009. Computer modelling of magnetron discharges, *J. Phys. D, Appl. Phys.*, 42:194018. DOI: 10.1088/0022-3727/42/19/194018. 30

[9] Baranov, O., Fang, J., Ostrikov, K., and Cvelbar, U. 2017. TiN deposition and morphology control by scalable plasma-assisted surface treatments, *Mater. Chem. Phys.*, 188:143. DOI: 10.1016/j.matchemphys.2016.12.010. 30

[10] Musil, J., Baroch, P., Vlcek, J., Nam, K. H., and Han, J. G. 2005. Reactive magnetron sputtering of thin films: Present status and trends, *Thin Solid Films*, 475(1–2):208. DOI: 10.1016/j.tsf.2004.07.041. 32

[11] Linss, V. 2014. Local racetrack plasma composition during reactive magnetron sputtering of ZnO:Al using rotatable Zn:Al target and influence on film properties, *Surf. Coat. Technol.*, 241:19. DOI: 10.1016/j.surfcoat.2013.09.053. 32, 37

[12] Baranov, O., Romanov, M., Wolter, M., Kumar, S., Zhong, X., and Ostrikov, K. 2010. Lowpressure planar magnetron discharge for surface deposition and nanofabrication, *Phys. Plasmas*, 17:053509. DOI: 10.1063/1.3431098. 32

[13] Honglertkongsakul, K., Chaiyakun, S., Witit-anun, N., Kongsri, W., and Limsuwan, P. 2012. Single langmuir probe measurements in an unbalanced magnetron sputtering system, *Procedia Eng.*, 32:962. DOI: 10.1016/j.proeng.2012.02.039. 32

[14] Svadkovski, I. V., Golosov, D. A., and Zavatskiy, S. M. 2003. Characterisation parameters for unbalanced magnetron sputtering systems, *Vacuum*, 68:283. DOI: 10.1016/s0042-207x(02)00385-8. 32

[15] Baranov, O., Romanov, M., Kumar, S., Zhong, X., and Ostrikov, K. 2011. Magnetic control of breakdown: Toward energy-efficient hollow-cathode magnetron discharges, *J. Appl. Phys.*, 109(6):063304. DOI: 10.1063/1.3553853. 32

[16] Kushner, M. J. 2003. Modeling of magnetically enhanced capacitively coupled plasma sources: Ar discharges, *J. Appl. Phys.*, 94(3):1436. DOI: 10.1063/1.1587887. 32

[17] Kim, D. W., Lee, H. Y., Kyoung, S. J., Kim, H. S., Sung, Y. J., Chae, S. H., and Yeom, G. Y. 2004. Magnetically enhanced inductively coupled plasma etching of 6H-SiC, *IEEE Trans. Plasma Sci.*, 32(3):1362. DOI: 10.1109/tps.2004.828821. 32

[18] Lieberman, M. A., Booth, J. P., Chabert, P., Rax, J. M., and Turner, M. M. 2002. *Plasma Sources Sci. Technol.*, 11:283. DOI: 10.1088/0963-0252/11/3/310. 32

[19] Perret, A., Chabert, P., Booth, J.-P., Jolly, J., Guillon, J., and Auvray, P. H. 2003. Ion flux nonuniformities in large-area high-frequency capacitive discharges, *Appl. Phys. Lett.*, 83(2):243. DOI: 10.1063/1.1592617. 32

[20] Meziani, T., Colpoand, P., and Rossi, F. 2001. Design of a magnetic-pole enhanced inductively coupled plasma source, *Plasma Sources Sci. Technol.* 10:276. DOI: 10.1088/0963-0252/10/2/317. 33

[21] Godyak, V. 2013. Ferromagnetic enhanced inductive plasma sources, *J. Phys. D, Appl. Phys.*, 46:283001. DOI: 10.1088/0022-3727/46/28/283001. 33

[22] Chen, F. F. 2015. Helicon discharges and sources: A review, *Plasma Sources Sci. Technol.*, 24:014001. DOI: 10.1088/0963-0252/24/1/014001. 33

[23] Chen, F. F. and Torreblanca, H. 2009. Permanent-magnet helicon sources and arrays: A new type of rf plasma, *Phys. Plasmas*, 16:057102. DOI: 10.1063/1.3089287. 34

[24] Uchida, T. and Hamaguchi, S. 2008. Magnetic neutral loop discharge (NLD) plasmas for surface processing, *J. Phys. D, Appl. Phys.*, 41:083001. DOI: 10.1088/0022-3727/41/8/083001. 34

[25] Sakurai, Y. and Osaga, T. 2011. Control of magnetic field in neutral loop discharge plasma for uniform distribution of ion flux on substrate, *IEEE Trans. Plasma Sci.*, 39(11):2550. DOI: 10.1109/tps.2011.2128895. 34

[26] Chen, F. 1984. *Introduction to Plasma Physics and Controlled Fusion*, New York, Plenum. DOI: 10.1007/978-1-4757-5595-4. 34

[27] Tang, D. and Chu, P. K. 2003. Current control for magnetized plasma in direct-current plasma-immersion ion implantation, *Appl. Phys. Lett.*, 82:2014. DOI: 10.1063/1.1564638. 34

[28] Yasaka, Y., Koga, K., Ishii, N., Yamamoto, T., Ando, M., and Takahashi, M. 2002. Planar microwave discharges with active control of plasma uniformity, *Phys. Plasmas*, 9(3):1029–1035. DOI: 10.1063/1.1447256. 34

[29] Yasaka, Y., Ishii, N., Yamamoto, T., Ando, M., and Takahashi, M. 2004. Development of a slot-excited planar microwave discharge device for uniform plasma processing, *IEEE Trans. Plasma Sci.*, 32(1):101. DOI: 10.1109/TPS.2004.823977. 34

[30] Yasaka, Y., Tobita, N., and Tsuji, A. 2013. Control of plasma profile in microwave discharges via inverse-problem approach, *AIP Adv.*, 3:122102. DOI: 10.1063/1.4840735. 34

[31] Bowles, J. H., Duncan, D., Walker, D. N., Amatucci, W. E., and Antoniades, J. A. 1996. A large volume microwave plasma source, *Rev. Sci. Instrum.*, 67(2):455. DOI: 10.1063/1.1146612. 34

[32] Fietzke, F., Morgner, H., and Gunther, S. 2009. Magnetically enhanced hollow cathode—a new plasma source for high-rate deposition processes, *Plasma Process. Polym.*, 6:S242–S246. DOI: 10.1002/ppap.200930607. 34

[33] Anders, A. and Brown, J. 2011. A plasma lens for magnetron sputtering, *IEEE Trans. Plasma Sci.*, 39(11):2528. DOI: 10.1109/tps.2011.2157172. 34

[34] Levchenko, I., Romanov, M., and Korobov, M. 2004. Plasma jet interaction with a spherical target in magnetic field, *IEEE Trans. Plasma Sci.*, 32(5):2139–2143. DOI: 10.1109/tps.2004.835527. 35

[35] Levchenko, I., Romanov, M., Baranov, O., and Keidar, M. 2003. Ion deposition in a crossed $E \times B$ field system with vacuum arc plasma sources, *Vacuum*, 72(3):335–344. DOI: 10.1016/j.vacuum.2003.09.002. 35

[36] Levchenko, I. and Baranov, O. 2003. Simulation of island behavior in discontinuous film growth, *Vacuum*, 72:205. DOI: 10.1016/j.vacuum.2003.08.004. 35

[37] Baranov, O. and Romanov, M. 2009. Process intensification in vacuum arc deposition setups plasma process, *Plasma Process. Polym.*, 6(2):95. DOI: 10.1002/ppap.200800131. 35

[38] Baranov, O., Romanov, M., Fang, J., Cvelbar, U., and Ostrikov, K. 2012. Control of ion density distribution by magnetic traps for plasma electrons, *J. Appl. Phys.*, 112(7):073302. DOI: 10.1063/1.4757022. 35, 37

[39] Baranov, O., Romanov, M., and Ostrikov, K. 2009. Effective control of ion fluxes over large areas by magnetic fields: From narrow beams to highly uniform fluxes, *Phys. Plasmas*, 16:053505. DOI: 10.1063/1.3130267. 35

[40] Davidson, R. C. 1976. Vlasov equilibrium and nonlocal stability properties of an inhomogeneous plasma column, *Phys. Fluids*, 19:1189. DOI: 10.1063/1.861601. 37

[41] Landau, L. D. and Lifshitz, E. M. 1976. *Mechanics*, Oxford, Butterworth–Heinemann. 37

[42] Baranov, O. O., Cvelbar, U., and Bazaka, K. 2018. Concept of a magnetically enhanced vacuum arc thruster with controlled distribution of ion flux, *IEEE Trans. Plasma Sci.*, 46(2):304–310. DOI: 10.1109/tps.2017.2778880. 37

CHAPTER 4

Material Processing

ABSTRACT

This chapter describes several fundamental processes in material treatment, such as etching, deposition, and modification, and how they can be used to enable different stages of nanofabrication. Specific examples that cover various technological approaches and materials are discussed and analyzed, presenting a wide range of modifications that can be achieved using these plasma-based tools.

4.1 TRADITIONAL PROCESSES: ETCHING, DEPOSITION, AND MODIFICATION

Plasma etching is a chemical reaction that takes place on the surface of the processed material in the presence of plasma. The most well-known application of the technology is the production of high aspect ratio (HAR) silicon, and a comprehensive review of the process was conducted by Wu et al. [1]. The necessity of the plasma etch development was spurred by the requirements implied by the fast growing branch of the microelectromechanical systems (MEMS), i.e., the production of microsensors, valves, probes, etc.

Anisotropic ion bombardment is one of the key points of the etching. It relies on the property of plasmas to create a region of the strong electric field between the quasineutral plasma and the electrically charged electrode immersed into the plasma. To enhance the sheath, either DC or CCP discharge may be ignited, and they both provide the means to control the ion energy under the condition of low gas pressure in the reactor. The latter requirement follows from the necessity of not losing the ion energy acquired in the sheath in the ion-neutral charge-exchange collision; for this reason, the sheath has to be collisionless. Generally, for the successful ion sputtering and etching, the plasma ions should gain the energy with a momentum directed perpendicular to the material surface during the path across the sheath. In a case of a complex surface relief, to meet this condition a matrix sheath should be developed by applying short pulses of a negative high-voltage bias to the substrate; CCP methods of control are also useful. To meet the specifications of the modern industry, a process called passivation is often used in conjunction with the ion bombardment. The most popular technology in HAR silicon etch is a time-multiplexed alternating process, also referred to as the Bosch method. This approach alternates sidewall passivation with etching; a schematic of the method is shown in Figure 4.1.

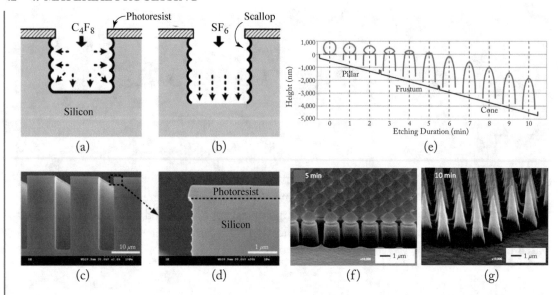

Figure 4.1: Schematic of the Bosch method. (a) Sidewall passivation is performed using C_4F_8, (b) silicon isotropic etching step employs SF_6 gas, (c) a SEM image of deep trenches formed as a result of this process, and (d) a magnified SEM image of the sidewall with nanoscallops. Reprinted with permission from Wu et al., *J. Appl. Phys.*, 108:051101, 2010. Copyright 2010 AIP. (e) Simulated results of the polymethyl methacrylate shape change by the use of a polystyrene etching mask. (f, g) SEM photographs of the array structures of pillars, frustums, and cones formed on polymethyl methacrylate plates. Reprinted with permission from Nabesawa et al., *AIP Adv.*, 3:112105 (2013). Copyright 2013 the Authors, under Creative Commons Attribution 3.0 Unported License.

As was stated, various reactors based on different plasma generation methods may be employed for surface etching; these include CCP [2–4], ECR [5], ICP [6, 7], ICP-CCP [8], RIE [9], and helicon plasma [10]. In the case of CCP, RF power with a typical frequency of 13.56 MHz is applied to the electrodes, and the manner in which power is delivered into the system determines the operation mode. When the power is applied to the electrode that holds the wafer with the workpieces, the mode is called the reactive ion etching (RIE) mode. In planar etch (PE) mode, the etched material is placed on the grounded electrode, and the etching effect is less severe than for the RIE mode. These modes can be combined in some plasma setups with the improved control of the etch mode. It should be noted that increased gas pressure provides for additional control over the ion energy, that the effect of the PE system may be achieved by implementation of the low-power and high-pressure RIE mode. ICP etch is also present in various modifications of the system. If the coil is placed above the dielectric window, the reactor is planar ICP; one more configuration of the setup is called a "transformer coupled plasma" reactor where the highly uniform plasma is achieved. When the coil is wound around a ceramic tube, the result is a tube ICP reactor. ECR etch reactors are also employed yet not to the same extent as the ICP reactors which are considered to be more reliable.

Coating deposition is a branch of plasma technology that withstood the test of time since the first experiments conducted at the end of the 19th century [11]. Generally, to get a deposit on a surface, one should employ species other than gas ions. For the purpose, magnetron sputtering and arc deposition are often the best choice of a deposition platform. However, both techniques have their advantages and disadvantages. A glow discharge modified by use of the magnetic field into the magnetron discharge has the advantage of abundant production of the ions derived from the background gas, while the fraction of the ions of the sputtered material represent only a few percent of the total flow of the sputtered neutrals [12]. Here, the sputtered flow is composed largely of an atomic fraction without the micro-droplets or clusters. However, to get the desired dense void-free condensates, plasma chemistry requires a much higher fraction of ions, and various ionized sputtering techniques have been developed to address this challenge and to achieve a degree of ionization of up to 50–90% for the sputtered atoms. Usually, these techniques rely on the introduction of the ICP power into the magnetron sputtering (ICP-MS); MW power can also be used in place of the ICP power, producing a microwave amplified magnetron sputtering system.

High performance of the sputtering process is achieved when high power pulses are supplied to the magnetron source at a particular duty cycle, which results in the development of high power impulse magnetron sputtering (HIPIMS), or implementation of the self-sustained sputtering (SSS), and the hollow cathode magnetron (HCM) sputtering [13].

Cathodic cage plasma deposition (CCPD) is a modification of the sputtering technique, when the treated samples are covered by a cage to avoid direct interaction with the plasma. The key benefits of using the CCPD technique include processing uniformity, and higher rate of film deposition at increased pressures, as well as reduced operation times; these benefits are well demonstrated when growing TiN and TiO_2 thin films on Si substrates using CCPD [14].

Vacuum arc coating technology [15] allows one to produce high-density, fully ionized fluxes of the evaporated material. In particular, it is widely applied in the production of reliable high quality TiN coatings for materials used in load bearing and biomedical applications [16]. Here, the metallic titanium evaporated from a cathode by use of the electric arc reacts with nitrogen introduced into the reactor to grow TiN films on a substrate exposed to the arc plasma. Vacuum arc deposition is a robust method with high productivity and well-developed methods of the process control based on changing the substrate bias and temperature, filling the reactor chamber with various gas mixtures at the different pressure, introducing the magnetic filters to remove the droplets and clusters from the plasma flux and others [17]. The above-discussed magnetic control of plasma discharge is widely applied in arc deposition, and its ability to control the chemical composition, preferred orientation, wear resistance, and hardness of TiN films has been confirmed by many experiments [18].

Figure 4.2: A trench with a high aspect ratio covered with a Pt deposit: cross-section SEM images of a cleaved Si wafer showing a trench with an aspect ratio of 22 (a); a close-up SEM image of the upper part of the trench showing the Si substrate, the thermally grown SiO_2 layer, and the ALD Pt layer (b); a close-up SEM image of the lower part of the trench (c); images (b) and (c) show scallops along the z-axis and line edge roughness along the y-axis. Reprinted with permission from Erkens et al., *J. Chem. Phys.*, 146:052818, 2017. Copyright 2017 AIP. SEM micrographs of TiN coatings fabricated using an ion arc deposition with zero substrate bias, deposition time 60 min, and a substrate positioned 24 cm from the arc: facing the arc ×30,000 (d); opposite to the arc-fracture surface ×12000 (e). Reprinted with permission from Sanchez et al., *Acta Materialia*, 53:4079–4086, 2005. Copyright 2005 Elsevier.

All the above-discussed methods are implemented in the low pressure environments (not exceeding a few hundred Pa). However, atmospheric pressure plasma (APP) setups operated at ambient pressure are typically employed for large area plasma processing of textiles [19], and fabrication of solar panels [20].

The etching and deposition methods can be combined to achieve impressive results. Figure 4.2a–c illustrates the effect of the plasma-assisted atomic layer deposition (ALD) of Pt using O_2 plasma [21]. It was reported that plasma-assisted highly conformal Pt films were deposited in trenches with high aspect ratio (of 3–34), which is confirmed by SEM inspection. The examples of SEM images of high surface-to-area ratio TiN coatings deposited over electrodes used for cardiac pacing are shown in Figure 4.2d,e. Implantable metals face a number of unique challenges, including the requirement for biocompatibility, excellent mechanical stability, especially if subject to load, controlled corrosion rates, catalytic activity and electrical conductivity, and low contact potential when used for stimulation [22, 23]. Surface chemistry and morphology at macro, micro, and nanoscales play an important role in defining the specific interactions between the implanted material and its environment, determining the dynamics of protein and cell fouling, bacterial colonization, and degradation [24, 25].

Plasma modification is widely used when it is necessary to change the physical and chemical properties of a surface without changing its geometrical size. Often, the modification stage accompanies material deposition to form a so-called "duplex treatment." Ion carburization [26], boriding [27], and nitriding [28] technologies have all been developed to advance traditional thermal chemical treatments that have been known and used for a very long time, and they have the undisputed advantage of being conducted without the need of maintaining the thermal equilibrium conditions. The time of plasma modification usually is much shorter than that needed for thermal modification of the material. For example, the results of application of nitrogen plasma for surface nitriding of AISI-4140 steel were studied by Shah et al. [29]. In the study, ICP nitriding of AISI-4140 steel was conducted for 2, 6, 10, and 14 h at -3 mbar, 100 W input power and the substrate temperature of 300°C. Results confirmed the formation of a nitrogen-infused ferrite phase (N-Fe) as a result of the incorporation of nitrogen into the iron lattice as an interstitial solid solution. A significant improvement (450 HV) in the surface hardness was obtained when the samples were exposed to the nitrogen plasma for 14 h at a filling pressure of 2 mbar. DC plasma is widely applied for the nitriding yet the direct action of the plasma ions can lead to the undesired consequences. In active screen plasma nitriding, the treated parts are surrounded by a metal screen to which a cathodic potential is applied. The object to be modified is kept at a floating potential or subjected to a small cathodic bias power. This electrical configuration results in plasma forming on the surface of the metal screen rather than on that of the treated object. The screen is heated by the plasma, with resulting radiation from the screen supplying the heat needed to bring the treated object to the required temperature [30]. Dong et al. employed the double glow (DG) plasma silvering with active screen plasma nitriding to develop stainless steel (SS) surfaces with prolonged antibacterial activity [31]. The Ag/N double doping of the SS surfaces was achieve by sequentially subjecting the surface to the aforementioned plasma alloying processes, first conducting the DG surface alloying with Ag followed by (b) active screen low temperature plasma nitriding (Figure 4.3). The nitrogen plasma treatment is also applied in nano-electronics to modify a surface of the vertical graphene nanostructures, as was shown by Lin et al. [32]. In another example, the tribological properties of TiAl were significantly improved by subjecting the material to a duplex treatment of plasma carburization on plasma alloyed TiAl with Nb [33]. The NbC and Nb_2C carbides and excessive carbon that are introduced into the material as a result of this treatment and the resulting thicker hardened surface layer are the main contributors to the enhanced tribological performance of TiAl. Treatment of pure Ti and TiAl6V4 titanium alloy using pulsed-DC plasma boriding in the Ar–BCl_3 atmosphere at 700–900°C [34] resulted in the formation of a compact titanium boride layer was formed under the surface, characterized by an extremely high value of 2800 $HK_{0.01}$. The duplex diffusional surface treatment has also been used to induce the formation of cubic boron nitride (c-BN) in addition to Fe_2B and iron nitrides in boronitride coating [35].

The use of oxygen plasma treatment is widespread due to the remarkable chemical activity of oxygen which can produce both positive and negative ions in plasmas [36]. The modification of surfaces of carbon nanotubes by O_2 plasma was successfully applied to modify the properties of MWCNT powders [37]. It was proposed that O_2 plasma ion bombardment introduces the defects into the structure of MWCNTs, and the ion beam reacts with the amorphous carbon

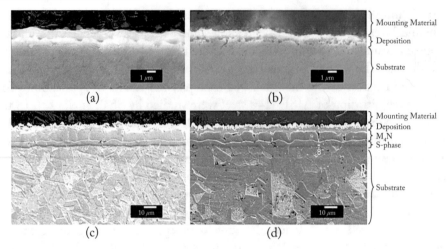

Figure 4.3: SEM images of the multilayer structure of treated stainless steels show deposition layers on (a) DG surface treatment without silver, (b) Ag–SS after the first step and multilayer structures of (c) DG surface treatment without silver with the following low temperature plasma nitriding, and (d) Ag—low-temperature plasma nitriding after the second step. Reprinted with permission from Dong et al., *Acta Biomater.*, 7(1):44–457, 2011. Copyright 2011 Elsevier.

until it is peeled from the nanotube. Highly effective oxygen plasma pretreatment was employed by Yang et al. to remove graphitic carbon from tungsten carbide surfaces without causing its oxidation to improve the interaction between metal adlayer modifiers and the WC substrate [38].

Experimental results have revealed that the plasma treatments could change and improve the surface properties of not only metals and alloys, but polymers [39]. Various types of plasma treatment could lead to surface modification or thin layer coating on the polymeric surface, while the control of both surface chemical composition and physical topography of the polymers is possible. Specifically, plasma surface modification and deposition can introduce functional groups, deliver surface roughness control, enable crosslink formation, graft polymerization, and deposit thin film coatings onto the polymeric surface [40]. Surface modification of polyester films by RF plasma was performed by Gupta et al. [41]. Plasma treatment of PET films was carried out under argon, followed by exposure to oxygen atmosphere. The films underwent a considerable change in surface composition and morphology, as demonstrated by contact angle measurements.

RF plasma was also used for chemical vapor deposition of thin polymer coatings from non-traditional precursors, e.g., plant-derived essential oils and their derivatives. Highly volatile at room temperature, these oils provide an interesting and potentially more sustainable alternative to conventional carbon sources [42]. When synthesized at low input power, the coatings derived from these essential oils often retain the useful antimicrobial properties of the precursor, with demonstrated ability to retard bacterial attachment as well as kill those pathogens that manage to attach [43, 44]. At the same time, they show excellent biocompatibility with tested mammalian cells lines, including epithelial and bone-forming cells. When deposited over biodegradable substrates, e.g., magnesium pins, these coatings can be used to control the degradation kinetics of the bulk metal [45]. Additionally, these thin films are mechanically flexible, optically transparent, and display an electron-blocking hole-transport property [46, 47], making them promising as components for thin film optoelectronics [48, 49].

4.2   PLASMA FOR NANOFABRICATION

Today, nano-world is a new frontier for the plasma fabrication [50]. For the past decade, various nanostructures have been developed, assembled, and processed using plasmas generated using a wide variety of systems [51]. For example, vertical graphene nanostructures have been successfully grown [52, 53] using DC glow [54, 55], CCP [56, 57], ICP [58, 59], MW [60], and ECR [61] plasmas. The growth mechanism of graphene in these systems have been extensively investigated to consider the key parameters of the plasma discharge [62], as well as the selection of the carbon source [63], and the choice of catalyst, although catalyst-free deposition has also been reported [64, 65]. Carbon nanotubes have been produced in arc [66], CCP [67], and ICP [68] reactors. A large diversity of nanostructures of different dimensionality (1D, 2D, and 3D) have been grown in oxygen plasmas [36, 69] with different plasma ignition mechanisms [70–72].

While different discharges can be used to synthesize a certain type of a nanoparticle, the effect of the discharge on the synthesis and deposition process and the properties of the final material may be quite different. A comparison of the use of ICP and PECVD to achieve the same task of preparation of amorphous and crystalline silicon interfaces on Si wafer was performed by Guo et al. [73], with the schematics of the setups shown in Figure 4.4. Here, 10-nm-thick amorphous a-Si:H thin films were deposited in the ICP-CVD system at a temperature of 100°C and 1300 W of RF power in the presence of such precursor gases as SiH_4 and H_2. At the second stage, the films were subjected to post-deposition plasma treatment in ICP or CCP discharges to improve the passivation effect. To do this, the thin films deposited onto n-type Si wafers and quartz substrates during the first stage were treated using hydrogen plasma in either the ICP-CVD or PECVD system.

In these setups, quite different operation modes are realized due to the differences in the plasma generation mechanisms. For the ICP discharge, the gas pressure was set at 2 Pa and the RF power was 1300–2700 W to reach the plasma density of 33.4–54.0 mW cm^{-3}. When using the PECVD discharge, the corresponding operation values were 133 Pa, 5–25 W, and 18.9–94.3 mW cm^{-3}, respectively. After the substrates were exposed to the plasmas for 2–4.5 min, the temperature of the substrate reached the value of 70–190°C. The experiments confirmed that the hydrogen content in the films increased as a result of the H treatment, and the increase was more significant after PECVD than after an ICP-based process. In Figure 4.5 the results of the

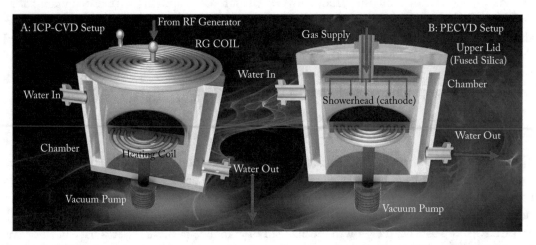

Figure 4.4: Two experimental setups used to form the amorphous/cr (460 kHz, 3.0 kW) (a) and PECVD where the wafer is equipped with a separate coil for the resistive heating (b). Reprinted with permission from Guo et al., *Appl. Surf. Sci.*, 427:486, 2018. Copyright 2018 Elsevier.

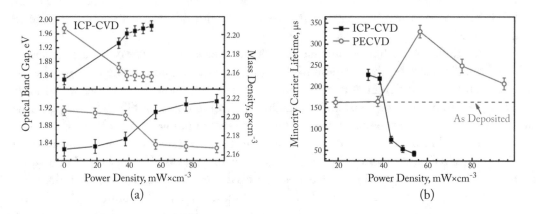

Figure 4.5: Effect of different plasmas on the characteristics of a-Si:H films: Dependence of the optical band gap and mass density on the power density of hydrogen plasma treatment in ICP-CVD and PECVD, where the black line represents the results for the optical band gap, while the red line shows results for the mass density (a); dependence of the minority carrier lifetime of a-Si:H/c-Si/a-Si:H heterostructures on the RF power during ICP-CVD and PECVD processing (b). Reprinted with permission from Guo et al., *Appl. Surf. Sci.*, 427:486, 2018. Copyright 2018 Elsevier.

measurement of the dependencies of optical band gap and mass density of the a-Si:H thin films on power density are shown. For both the ICP and CCP hydrogen plasma treatment (HPT), the optical band gap increased with the power, and the mass density monotonically decreased with the power density increase.

Here, we can see the difference in the power applied to obtain the results in both reactors. Specifically, the results for the effective minority carrier lifetime of a-Si:H/c-Si/a-Si:H heterostructures as a function of power density in ICP-CVD and CCP discharges differ significantly. For ICP, the carrier lifetime decreased monotonically as the power density increased, thus indicating the crystallization at the interface. For the PECVD treatment, there was no obvious improvement in the carrier lifetime. Hence, at low power densities in PECVD plasmas, the hydrogen plasma was not able to deliver sufficient energy for hydrogen diffusion to the interface to improve passivation. In addition, the number density of micro-voids was much higher after the ICP treatment.

CCP pre-treatment was also proposed and successfully tested to reduce the water vapor transmission rate (WVTR) and water contact angle of thin polyethylene terephthalate (PET) with a deposited alumina film [74]. Here, the researchers needed to deal with the necessity of sustaining low temperatures of about 50–110°C to avoid the detrimental effect of the differences in thermal expansion between the PET and Al_2O_3 films that could compromise the interlayer adhesion, an at the same time enable good adhesion of the film to the substrate that generally requires the use of higher temperatures. Oxygen plasma was employed in this case to create more polar groups (such as carbonyl and hydroxyl) on and under the polymer surface, promoting the interfacial reactions [75]. In addition, oxygen plasma creates a greater number of oxygen-contained radicals when compared to the inert gas plasma, and hence, ensures the formation of a greater number of polar groups and superior moisture barrier performance [76].

In the experiment shown in Figure 4.6a, the technology based on plasma-enhanced hot filament deposition was used to synthesize a composition of vertically oriented graphene nanoflakes/BNCO nanowalls. A very high level of plasma control was needed here to create this complex hierarchical structure [77]. A similar technique was used to synthesize carbon nanoflakes on gold nanodots in plasma [78].

CP treatment has been successfully employed to change mechanical characteristics of a surface, such as its roughness that can limit the performance of Si-based solar cells. The results shown in Figure 4.7 demonstrate that a desirable level of roughness can be attained following a simple exposure of p-type Si (100) wafers to low-temperature ICP plasmas generated using mixtures of argon and hydrogen [79]. A schematic of the experimental setup is depicted in Figure 4.7a, and a photo of the discharge generated in this system is presented in Figure 4.7b. The schematic of the process and the SEM visualization of the resulting Si nanostructure arrays are shown in Figures 4.7c,d. Four typical morphologies obtained using different hydrogen/argon ratios are shown in Figures 4.7e–h. One can see that the pure argon plasma did not produce well-resolved nanostructures (morphology 1); in this case, the surface produces a very weak

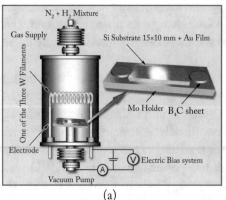
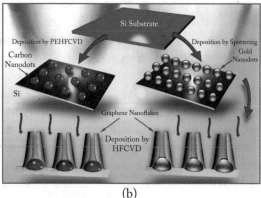

(a)   (b)

Figure 4.6: (a) Schematic of the experimental plasma-enhanced hot filament chemical vapor deposition system used to synthesize graphene nanoflake/BNCO composition using a B_4C precursor. N_2 and H_2 were used as reactive gases. The bias current in the electric bias system was gradually increased until a blue glow appeared near the substrate, and then it was set to 160 mA to grow the nanowalls. Reprinted with permission from Wang et al., *J. Mater. Chem. C*, 4(2):9788–9797, 2016. Copyright 20167 RSC. (b) Schematic diagram of plasma-assisted growth of carbon nanoflakes. Deposition by PEHFCVD and by sputtering was used to form arrays of carbon and gold nanodots on Si substrates. Then, deposition by HFCVD was used to nucleate and grow vertically aligned carbon nano flakes on both carbon and gold nanodots. Reprinted with permission from Wang et al., *Carbon*, 124:391–402. Copyright 2017 Elsevier.

photovoltaic (PV) response, not exceeding 0.49%. However, when hydrogen is introduced into the mixture at a ratio $H_2/Ar = 0.1$, large nanopyramid-like nanostructures are visible on the surface (morphology 2); the PV response is also stronger (1.44%). Increasing the content of hydrogen in the mixture to $H_2/Ar = 0.33$ results in a nanoarray of sharp nanocones being produced (morphology 3), with the PV response reaching 5.1%. Finally, the mixture of $H_2/Ar = 0.08$ results in a two-level hierarchical array of nanocones (morphology 4); this treatment results in the solar cell efficiency of up to 11.9%, fill factor of up to 70%, and a short-circuit current of up to 32 mA/cm^2.

In the treatment, the plasma particles may play a number of different roles: electrons cause heating of the Si surface, the argon ions are responsible for the sputtering and roughening of the wafer, and the reactive hydrogen ions and radicals etch the Si surface by forming volatile Si-containing molecules of various complexity. Along with other variables, the ratio of argon to hydrogen determines the final morphology of the substrate surface.

In a study conducted by Primč et al., [80], a widely used thermal CVD furnace was modified by adding two cathodes and anode, which allows one to ignite the glow discharge within two distinct regions. One is a long (~ 0.5 m) discharge within the heated furnace zone, and an-

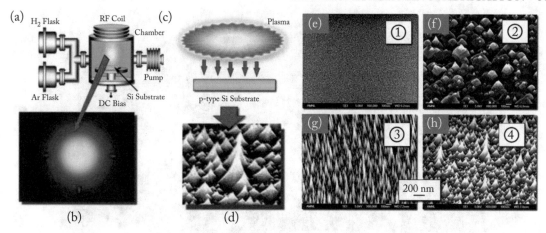

Figure 4.7: ICP H$_2$+Ar treatment to improve the performance of Si-based solar cells: a schematic of the experimental setup (a); a photograph of the plasma glow (b); a scheme of the deposition process (c); a representative SEM image of a Si nanostructure array (d); typical SEM images of topographies produced using different levels of hydrogen in the gas mixture: H$_2$/Ar = 0, weak PV response (e); H$_2$/Ar = 0.11, stronger PV response (f); H$_2$/Ar = 0.33, good PV response (g); H$_2$/Ar = 0.08, the best PV response (h). All patterns were grown at the discharge power of 2 kW, substrate bias $U_s = -50$ V, substrate temperature $T_s = 500°$C, process time of 30 min. Reprinted with permission from Xu et al., *Adv. Energy Mater.*, 1:373–376, 2011. Copyright 2011, Willey.

other is a shorter (~ 0.2 m) discharge which is generated outside the heated zone. Figures 4.8a,b shows the schematic of these two distinct discharge configurations, and Figures 4.8c,d shows the discharges produced using a mixture of Ar and O$_2$ gas at the gas pressure of about 1 mbar. The study also demonstrated that is possible to use this PECVD setup to grow hierarchical ZnO nanowires that can be seen in Figure 4.8e. Figure 4.8f depicts the entire deposition system.

A synchronous ICP discharge in the N$_2$–H$_2$ gas mixture was used to develop stress sensors based on the graphene oxide, with the view to use these sensors to monitor various parameters in a human body [81]. The applied RF power of 1400 W produced the high-density plasma in the pressure range of a few mTorr, with the system allowing one to independently control ion energy and their density. In addition, the plasma-induced electric field was controlled using a negative DC bias to further increase the temperature of electrons.

Figure 4.9 illustrates the plasma process used to concomitantly reduce and introduce nitrogen doping into graphene oxide deposited on polydimethylsiloxane substrates. The strongly chemically reactive, highly dense plasma environment rich in reactive H radicals and N$_2^+$ ions promoted the efficient reduction and doping of GO. Here, the H radicals play a role of a reactant, removing oxygen-containing functional groups from GO, whereas high-energy ion colli-

Figure 4.8: Long- and short-type glow discharges produced in the modified CVD reactor: schematics of a short discharge in the cold zone (a) and a long discharge in the hot zone (b); photos of the plasma generated in the short (c) and long discharges (d). In (d), -the cathode glow is clearly separated from the positive column by the cathode dark space; a representative SEM image of the hierarchical ZnO nanowires grown in the reactor (e); a general view of the PECVD system with the discharge ignited in the long configuration (f). Reprinted with permission from Primč et al., *IEEE Trans. Plasma Sci.*, 42(10):2564–2565, 2014. Copyright 2014 IEEE.

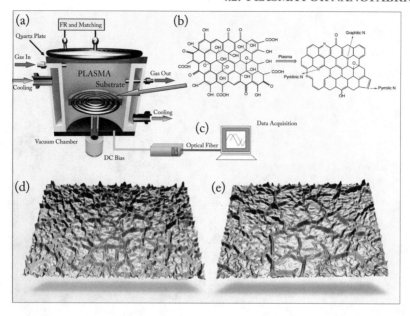

Figure 4.9: N_2–H_2 ICP conversion of graphene oxide (GO) to nitrogen-doped reduced GO (N-RGO): a schematic of the experimental setup (a); a schematic of the process of nitrogen doping and reduction of GO (b); OES measurement system used to characterize the plasma (c); three-dimensional reliefs of the original graphene oxide (d) and nitrogen-doped and reduced GO (e). Reprinted with permission from Zhou et al., *ACS Appl. Mater. Interfaces*, 11:15122–15132, 2019. Copyright 2019 ACS.

sions introduce Ni into the GO structure. The pathway for N_2–H_2 plasma-enabled conversion of graphene oxide (GO) into the nitrogen-doped reduced GOs (N-RGO) is shown in Figure 4.9b.

Thus-prepared N-RGO films were treated under the negative bias of either 0 or 35 V. Surfaces of plasma-processed N-RGO-35V films were found to be smoother, thus indicating that the number of defects in the graphene oxide were reduced as a result of plasma processing. This effect is attributed to the enhanced ion bombardment characterized by deeper penetration of the ions into the GO film. Biasing the substrate also increased the level of pyridinic N and graphitic N in the GO film.

Schematics of the two typical ICP reactors for the growth of various surface-bound nanostructures such as nanorods, nanocones, nanowires, graphene, and large patterns of nanodots are shown in Figure 4.10 [82]. A typical reactor configuration includes a vacuum chamber fitted with a planar inductive coil mounted on its upper flange, as shown in Figure 4.10a. Alternatively, the coil can be mounted over the confinement tube, as shown in Figure 4.10b. One should note the ICP operating in the H-mode is typically characterized by a significantly greater plasma density

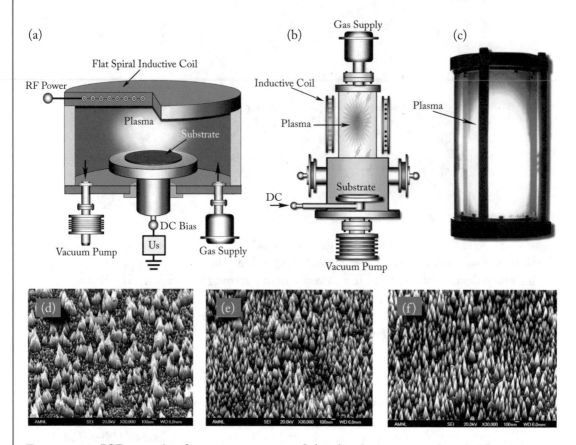

Figure 4.10: ICP growth of nanocones on a n-Si(100) substrate coated with Ni catalyst: a schematic of the ICP reactor with a planar coil mounted on top of the vacuum chamber (a); a schematic of the ICP reactor with the coil installed over the confinement chamber (b); a photo of the discharge (c) in the confinement chamber of the setup shown in (b). Reprinted with permission from Levchenko et al., *J. Vac. Sci. Technol. B*, 31(5), 2013. Copyright 2013 AIP. SEM images of carbon nanocones grown in an ICP reactor for different growth times: (a) 5 min; (b) 10 min; and (c) 20 min. Reprinted with permission from Tsakadze et al., *Carbon*, 45:2022, 2007. Copyright 2007, Elsevier Ltd.

than that in the CCP; a photograph of the ICP discharge generated in a typical reactor is shown in Figure 4.10c.

In one example, the nanocones were fabricated on a lightly doped n-Si(100) wafer pre-coated with a 30-nm-thick layer of Ni catalyst using magnetron sputtering prior to placing the substrates into the ICP reactor. Once the samples were installed in the reactor, and pre-treated by the argon discharge for 30 min [83]. Following this pre-treatment, a discharge was ignited in a mixture of hydrogen and argon to prepare the catalyst layer for nucleation, where the metal layer was fragmented into separate islands. Then methane was introduced into the reactor, and the nanocones began to grow. In this stage of fabrication, the growing nanostructures were heated by the plasma, and the surface temperature of about 500°C was reached at the negative biasing potential of 300 V. The RF power of 0.09–0.11 W×cm-1 was used to sustain the density of the plasma of 10^{12} cm^{-3}. The deposition time of 5 min produced a highly non-uniform pattern, whereas extending the deposition time to 10 min significantly improved the uniformity of the pattern, with patterns with good uniformity and significantly thinner well-defined nanocones produced after 20 min of the treatment (Figures 4.10d–f).

The ICP post-treatment to induce the formation of graphenes on the surface of alumina (Al_2O_3) membranes prepared by liquid anodization was conducted by Fang et al., [84], as shown in Figure 4.11. After the anodization of the Al foil, the samples were exposed to different post-treatment regiments, with one set of samples being subjected to an ICP plasma treatment in atmosphere of Ar gas with an RF power of 750 W applied for 2 min. The second set of samples had a 30 nm-thick layer of gold deposited on their surface, while the others were subjected to Au, CO, and Pd ions by immersing the samples into various chemical liquid reagents for 24 hr. This was followed by processing the samples in an ICP reactor in a $CH_4/Ar/H_2$ (2:1:1) mixture at the gas pressure of 3.0 Pa, RF power of 850 W, and process time of 6 min. This stage resulted in the formation of a mesh-like network of vertically-aligned graphene flakes on all samples. In a separate parallel experiment, the few-layered graphene flakes were deposited on thick Si(100) substrates covered with a 500 nm silica (SO_2) layer. These samples were first subjected to a treatment of low-temperature Ar plasma, and then transferred to the ICP reactor to facilitate the catalyst-free growth of graphene flakes under similar fabrication conditions.

A negative bias was applied to the substrate to control the structure of thus-produced graphenes, with −50 V potential producing petal-like structures when the gas mixture contained a greater fraction of H_2 and a lower content of CH_4. In contrast, increasing the fraction of CH_4 and reducing the H_2 content, as well as changing the bias to −100 V produced tree-like graphene networks. The temperature of the substrates reached 700–750 K as a result of plasma treatment.

A modification of the process shown in Figure 4.12 was reported by the same research group [85]. Here, some of the samples were coated with a 5-nm-thick gold layer deposited using a plasma-assisted sputtering process after the first stage of anodization, when the nanoporous aluminum membranes were formed. To keep the pores open, the substrates were not rotated during the coating. Then all the samples were subjected to the ICP treatment for the nanowall

(a)   (b)   (c)

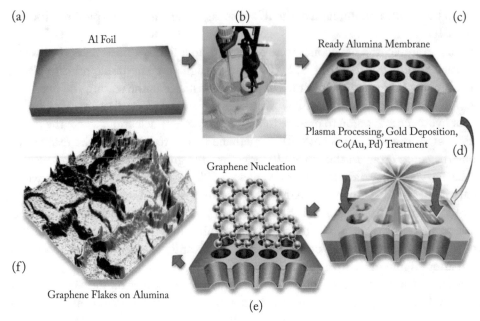

Al Foil

Ready Alumina Membrane

Plasma Processing, Gold Deposition,
Co(Au, Pd) Treatment

(d)

Graphene Nucleation

(f)

Graphene Flakes on Alumina

(e)

Figure 4.11: The use of the ICP discharge for graphene flake network fabrication on nanoporous alumina membranes. Aluminum foil (a) is subjected to an electrochemical treatment (b) to form the nanoporous membrane (c), which is then immersed into the plasma or chemical reagents (d), followed by the nucleation of the few-layered graphene flakes (e) and their growth to form a complex pattern of vertically-aligned nanowalls (f). Reprinted with permission from Fang et al., *Sci. Technol. Adv. Mater.*, 15:045004, 2014. Copyright 2016.

growth at an operational frequency of 13.56 MHz and the discharge power of 1.0 kW in an atmosphere of methane and argon. The stages of the process are shown schematically in Figure 4.12a, a schematic of the ICP reactor is shown in Figure 4.12b, SEM images of the original and gold-coated membranes top views are shown in Figures 4.12c,d. These structures could be used for many important applications requiring highly ordered porous materials, such as sensors [86, 87], filters [88, 89], and biosensors [90, 91].

An arc discharge method proved to be suitable for large scale production of graphene materials. High pressures in a range of 0.5–1 atm were used in the synthesis of the few-layer graphenes through the arc ignition between the carbon rods of 5–10 mm in diameter [92], as shown in Figure 4.13. In the process, a hole can be made in the center of the anode, which is then filled with the carbon catalyst, Y-Ni powder for example. The typical time required for the growth of graphenes was 1 ms, estimated from the characteristic velocities (10–100 m·s^{-1}) in the arc plasma and the width of the inter-electrode gap (several mm). The reported process offers a favorable combination of a larger size of graphene flakes, and very low density of adatom

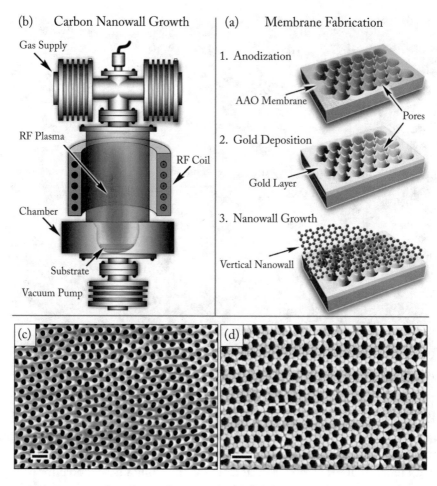

Figure 4.12: Fabrication ofgraphene-functionalized AAO membranes using an ICP treatment: three stages of the process (a); a schematic of the ICP plasma reactor used to grow carbon nanowalls (b); representative SEM images of the AAO membrane (60 nm pore size) prior to gold deposition (c) and after the deposition of the 5 nm gold layer (d). Scale bars are 200 nm. Reprinted with permission from Fang et al., *Carbon*, 78:627–632, 2014. Copyright 2014 Elsevier.

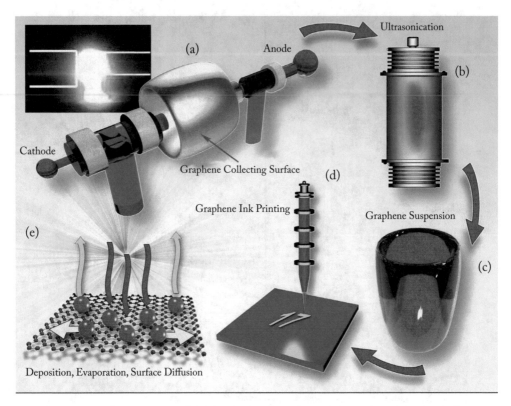

Figure 4.13: An arc-discharge method for the preparation of a few-layer graphene flakes in He gas atmosphere: schematics of the setup consisting of an anode, a cathode, a graphene collecting surface and (optionally) a focusing magnet. Inset shows the photo of the discharge produced in the system (a); removal of the carbon deposit from the collecting surface and ultrasonication (b); production of the graphene flake suspension (c); employment of the graphene ink for printing (d); and a schematic of the processes considered in the numerical simulation (e): an influx of the carbon atoms from the plasma to the growing graphene flake, the evaporation of carbon atoms from the graphene surface, diffusion of the carbon atoms along the graphene surface, and incorporation of carbon adatoms into the edges of the flakes. Reprinted with permission from Levchenko et al., *Graphene*, 5:81–89, 2016. Copyright 2016, Authors.

on the surfaces of thus-grown graphene, where the latter promotes the growth of single-layered flakes.

Numerical studies conducted with a purpose to reveal the growth mechanisms of graphenes in arc plasmas considered three key family of processes: (1) the influx of carbon atoms from the plasma to the surface of the graphene flake as it grows; (2) the evaporation of adsorbed atoms (adatoms) from the surface of graphene flakes into the plasma; and (3) surface diffusion of the atoms along the surface of graphenes, and their subsequent incorporation into the structure of the graphene flake at its edges; the latter enables the growth of the flake. The results show that the activation energies of the surface diffusion ε_d and carbon evaporation ε_a strongly influence the growth process. According to the calculations, the density of carbon adatoms is very low ($\approx 10^{-3}$ of monolayer density) when the temperature of a graphene flake (growth temperature) is 1300 K (when the flake is growing in the arc plasma jet), effectively reducing the probability of nucleation of new graphene layers. On the other hand, cooling of the flake increases the density of the adatoms; the temperature changes from 1300–1000 K when the flake leaves the plasma jet. During the cooling stage, the relative number density of the adatoms reaches the value of 0.3 (i.e., one adatom per three surface atoms); these conditions are more favorable for the nucleation and subsequent transition from a single- to a two-layer graphene flake structure. The temperature of electrons in the plasma strongly affects the adatom density , attributed to the effect on the Bohm velocity. For example, plasmas with the electron energies of 0.5 and 1.1 eV produces flakes with the sizes that vary by the factor of 2, while the respective adatom density on the flakes differs by the factor of 3. It is possible that higher temperatures may lead to increased desorption of adatoms from the graphene surfaces.

Unsupported graphene flakes were fabricated using an arc discharge-assisted system depicted in Figure 4.14c,d. Here, a magnetic unit is used to enhance the discharge and enable the separation to collect thus-produced carbon nanostructures [93]. In this system, the temperature of the anode can reach 3500 K, with the intense erosion of the electrode material in the discharge spots resulting in the production of a dense thermal-equilibrium plasma. The temperature of the gas in the discharge can reach 5000 K; as such, it is possible to effectively heat the graphene flakes to the temperatures needed to sustain efficient graphene formation even if the flakes swiftly transit through the reaction zone at a very high speed (with a typical residence time in the reaction zone of 10^{-4}s).

An external magnetic coil or a magnet mounted above the processing area produces a magnetic field that can significantly enhance the density of the plasma via magnetic confinement by trapping the plasma in the vicinity of the electrodes, as well as through the magnetization of plasma electrons which enhances the ionization of the neutral atoms. The electric field, which is stronger in the magnetized plasma, also result in the increased temperature of the electrons, promoting graphene heating in the plasma and sustaining an intense flux of carbon material to the graphene surfaces. Jointly, the aforementioned effects contribute to the efficient nucleation, rapid growth and greater crystallinity of the graphene flakes. As a result of the process, carbon

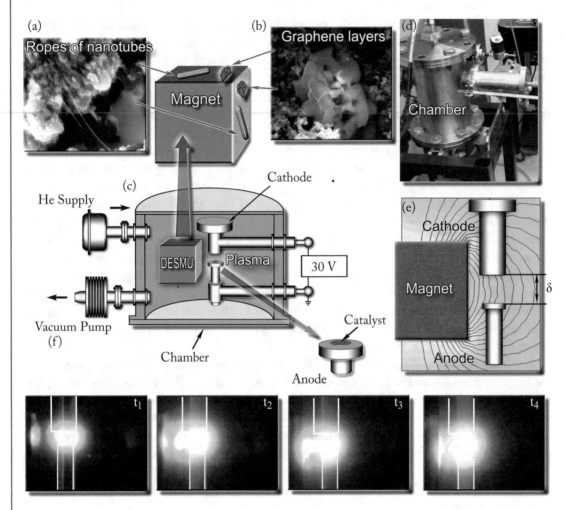

Figure 4.14: An arc discharge reactor with a magnetic unit used to collect the carbon nanostructures produced in the arc plasma: SEM images of the carbon deposit collected from the areas of the magnetic unit, which are close to the discharge (a), and far from the discharge (b); a schematic of the reactor (c); a photograph of the experimental setup (d); a computed 2D map of the magnetic field (e); and consecutive photographs of the discharge (f). Reprinted with permission from Volotskova et al., *Nanoscale*, 2:2281, 2010. Copyright 2010 RSC.

nanotubes are grown on the top and side surfaces in the areas of the magnetic unit which are close to the discharge (Figure 4.14a); graphene layers are found on the top and side surfaces of the unit, which are remote from the discharge (Figure 4.14b); thus, an effective separation of the two different types of the carbon nanostructures was ensured by the magnetic field of 0.12 T in the discharge gap (Figure 4.14e). Time evolution of the discharge is shown in a series of the photographs Figure 4.14f: the discharge is localized between the electrodes at the first moments after ignition (t_1); then plasma expands toward the magnetic unit (t_2, t_3, and t_4).

A transient spark (TS) discharge can be generated at an atmospheric pressure when using noble gases such as helium, argon and their mixtures in a pressure-jet configuration depicted in Figure 4.15a to produce Mo nanoflakes [94]. Testing of different reactor configurations that can be used to generate atmospheric-pressure discharges showed the possibility of fabricating several different types of nanostructures, including Mo nanoflakes. Figure 4.15b shows an RF dielectric barrier discharge (DBD) produced by supplying the power of 25 W at 1.9 kV and a frequency of 350-kHz. The discharge is generated within the glass tube containing a pin electrode, with a ring electrode fitted around its outer surface, in the proximity of the exit. It is possible to generate a larger discharge inside a glass vessel with a diameter of 5 cm, filling the tube with plasma, by modifying the pin electrode to have a spiral shape, and installing the external electrode under the glass vessel, as shown in Figure 4.15c.

Alternatively, the RF DBD can be generated using a tubular central electrode, as pictures in Figure 4.15d. SEM images of Mo nanostructures (flowerlike and spherical) synthesized on the Cu substrate by means of the nanosecond repetitively pulsed (NRP) spark discharge in ambient air are shown in Figures 4.15e,f. Flowerlike Mo nanoflakes that are generated when either a pin–pin or pin–plane configuration is used.

For applications requiring the growth of large arrays of vertically aligned carbon nanowalls, microwave plasmas are often used [95]. Surface-wave-driven MW plasmas are also a powerful tool to synthesize other types of carbon nanostructures, as was shown in the experiments conducted by Bundaleska et al., [96]. In their setup, a MW generator operates at a frequency 2.45 GHz with the power of up to 2 kW, which allows effectively igniting argon-methane gas mixture. Complex processes that occur in the plasma reactor are depicted in a schematic shown in Figure 4.16. The first zone incorporates the zones inside the launcher and the hot plasma zone outside it, where the power is absorbed by the plasma electrons, which transfer the power to neutrals via collisions. As a result, the gas temperature reaches the value of up to 3000 K; this temperature is maintained at a nearly constant level up to the distance of $\sim$ 10 cm from the launcher, then dropping sharply in the afterglow plasma zone. In the hot plasma zone, the chemical reactions take place. The next zone is located at a distance of up to $\sim$ 20 cm in the axial direction; this zone plays an important role because it is a place where the growth and assembly of carbon nanostructures take place.

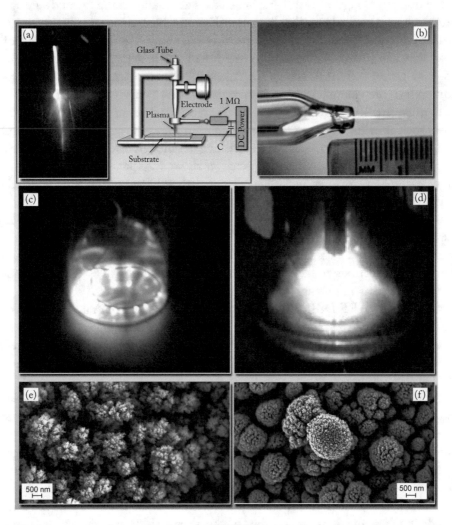

Figure 4.15: A nanosecond repetitively pulsed (NRP) spark discharge used to grow Mo nanostructures: a photograph of a transient spark discharge (a); an RF DBD discharge in a configuration of the atmospheric-pressure plasma jet (b); an atmospheric RF DBD generated in a glass vessel (c) and with the help of a tubular central electrode (d); SEM images of flowerlike Mo nanoflakes (e); and spherical Mo nanoflakes (f) produced using NRP spark discharges in ambient air. Reprinted with permission from Pai et al., *Plasma Sources Sci. Technol.*, 19:065015, 2011. Copyright 2011 IEEE.

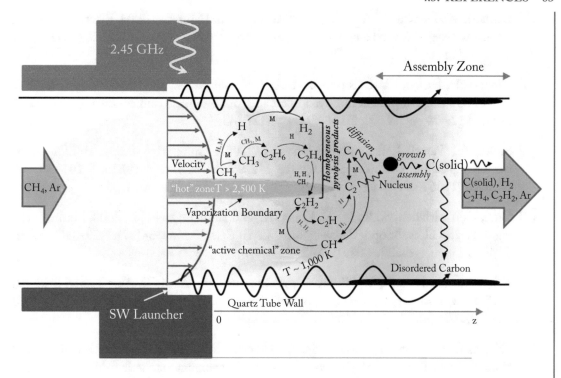

Figure 4.16: Scheme of the main processes in the microwave plasma. Reprinted with permission from Bundaleska et al., *Phys. Chem. Chem. Phys.*, 20:13810–13824, 2018. Copyright 2018 RSC.

4.3   REFERENCES

[1] Wu, B., Kumar, A., and Pamarthy, S. 2010. High aspect ratio silicon etch: A review, *J. Appl. Phys.*, 108:051101. DOI: 10.1063/1.3474652. 41

[2] Yen, T. F., Chang, K. J., and Chiu, K.-F. 2005. Profile control in high aspect ratio contact hole etching by a capacitively coupled plasma source, *Microelectron Eng.*, 82(2):129–135. DOI: 10.1016/j.mee.2005.07.001. 42

[3] Gul, B., Ahmad, I., Zia, G., and Aman-ur-Rehman 2016. Numerical study of capacitive coupled HBr/Cl2 plasma discharge for dry etch applications, *Phys. Plasmas*, 23(9):093508. DOI: 10.1063/1.4962570. 42

[4] Alshaltami, K. A. and Daniels, S. 2019. Investigation of etching optimization in capacitively coupled SF6-O2 plasma, *AIP Adv.*, 9(3):035047. DOI: 10.1063/1.5066286. 42

[5] Bernard, M., Deneuville, A., Ortega, L., Ayadi, K., and Muret, P. 2004. Electron cyclotron resonance oxygen plasma etching of diamond, *Diam. Relat. Mater.*, 13(2):287–291. DOI: 10.1016/j.diamond.2003.10.011. 42

[6] Shah, A. P., Laskar, M. R., Rahman, A. A., Gokhale, M. R., and Bhattacharya, A. 2013. Inductively coupled plasma reactive ion etching of III-nitride semiconductors, *AIP Conf. Proc.*, 1512:494–495. DOI: 10.1063/1.4791127. 42

[7] Stoltz, A., Varesi, J. B., and Benson, J. D. 2007. Comparing ICP and ECR etching of HgCdTe, CdZnTe, and CdTe, *J. Electron. Mater.*, 36(8):1007–1012. DOI: 10.1007/s11664-007-0163-z. 42

[8] Choi, D.-Y., Maden, S., Rode, A., Wang, R., and Luther-Davies, B. 2008. Plasma etching of As2S3 films for optical waveguides, *J. Non Cryst. Solids*, 354(27):3179–3183. DOI: 10.1016/j.jnoncrysol.2008.01.014. 42

[9] Nabesawa, H., Hiruma, T., Hitobo, T., Wakabayashi, S., Asaji, T., Abe, T., and Seki, M. 2013. Lowpressure plasma-etching of bulk polymer materials using gas mixture of CF4 and O2, *AIP Adv.*, 3(11):112105. DOI: 10.1063/1.4830277. 42

[10] Takahashi, K., Motomura, T., Ando, A., Kasashima, Y., Kikunaga, K., Uesugi, F., and Hara, S. 2014. Transport of a helicon plasma by a convergent magnetic field for high speed and compact plasma etching, *J. Phys. D, Appl. Phys.*, 47(42):425201. DOI: 10.1088/0022-3727/47/42/425201. 42

[11] Anders, A. 2002. From plasma immersion ion implantation to deposition: A historical perspective on principles and trends, *Surf. Coat. Technol.*, 156(1–3):3–12. DOI: 10.1016/s0257-8972(02)00066-x. 43

[12] Baranov, O., Romanov, M., and Ostrikov, K. 2009. Discharge parameters and dominant electron conductivity mechanism in a low-pressure planar magnetron discharge, *Phys. Plasmas*, 16:063505. DOI: 10.1063/1.3153554. 43

[13] Helmersson, U., Lattemann, M., Bohlmark, J., Ehiasarian, A. P., and Gudmundsson, J. T. 2006. Ionized physical vapor deposition (IPVD): A review of technology and applications, *Thin Solid Films*, 513(1–2):1–24. DOI: 10.1016/j.tsf.2006.03.033. 43

[14] De Sousa, R. R. M., Sato, P. S., Viana, B. C., Alves, C., Nishimoto, A., and Nascente, P. A. P. 2015. Cathodic cage plasma deposition of TiN and TiO2 thin films on silicon substrates, *J. Vac. Sci. Technol. A*, 33(4):041502. DOI: 10.1116/1.4919770. 43

[15] Anders, A. 2008. *Cathodic Arcs: From Fractal Spots to Energetic Condensation*, New York, Springer. DOI: 10.1007/978-0-387-79108-1. 43

[16] Baranov, O. O., Fang, J., Rider, A. E., Kumar, S., and Ostrikov, K. 2013. Effect of ion current density on the properties of vacuum arc-deposited TiN coatings, *IEEE Trans. Plasma Sci.*, 41(12):3640–3644. DOI: 10.1109/tps.2013.2286405. 43

[17] Sánchez, G., Rodrigo, A., and Alles, A. B. 2005. Titanium nitride pacing electrodes with high surface-to-area ratios, *Acta Mater.*, 53(15):4079–4086. DOI: 10.1016/j.actamat.2005.05.008. 43

[18] Zhao, Y.-H., Yang, W.-J., Guo, C.-Q., Chen, Y.-Q., Yu, B.-H., and Xiao, J.-Q. 2015. Effect of axial magnetic field on the microstructure, hardness and wear resistance of TiN films deposited by arc ion plating, *Acta Metall. Sin.-Engl.*, 28(8):984–993. DOI: 10.1007/s40195-015-0285-x. 43

[19] Cvelbar, U., Walsh, J. L., Černák, M., de Vries, H. W., Reuter, S., Belmonte, T. et al. 2018. White paper on the future of plasma science and technology in plastics and textiles, *Plasma Process. Polym.*, 16(1):1700228. DOI: 10.1002/ppap.201700228. 44

[20] Anand, V., Nair, A. R., Shivashankar, S. A., and Mohan Rao, G. 2015. Atmospheric pressure plasma chemical vapor deposition reactor for 100 mm wafers, optimized for minimum contamination at low gas flow rates, *Appl. Phys. Lett.*, 107(9):094103. DOI: 10.1063/1.4929781. 44

[21] Erkens, I. J. M., Verheijen, M. A., Knoops, H. C. M., Keuning, W., Roozeboom, F., and Kessels, W. M. M. 2017. Plasma-assisted atomic layer deposition of conformal Pt films in high aspect ratio trenches, *J. Chem. Phys.*, 146(5):052818. DOI: 10.1063/1.4972120. 44

[22] Bazaka, K. and Jacob, M. V. 2013. Implantable devices: Issues and challenges, *Electronics*, 2(1):1–34. DOI: 10.3390/electronics2010001. 44

[23] Prasad, K., Bazaka, O., Chua, M., Rochford, M., Fedrick, L., Spoor, J. et al. 2017. Metallic biomaterials: Current challenges and opportunities, *Materials*, 10(8):E884. DOI: 10.3390/ma10080884. 44

[24] Bazaka, K., Jacob, M. V., Crawford, R. J., and Ivanova, E. P. 2012. Efficient surface modification of biomaterial to prevent biofilm formation and the attachment of microorganisms, *Appl. Microbiol. Biotechnol.*, 95(2):299–311. DOI: 10.1007/s00253-012-4144-7. 44

[25] Bazaka, K., Crawford, R. J., and Ivanova, E. P. 2011. Do bacteria differentiate between degrees of nanoscale surface roughness? *Biotechnol. J.*, 6(9):1103–1114. DOI: 10.1002/biot.201100027. 44

[26] Edenhofer, B., Gräfen, W., and Müller-Ziller, J. 2001. Plasma-carburising—a surface heat treatment process for the new century, *Surf. Coat. Technol.*, pp. 142–144, pp. 225–234. DOI: 10.1016/s0257-8972(01)01136-7. 44

[27] Filep, E. and Farkas, S. 2005. Kinetics of plasma-assisted boriding, *Surf. Coat. Technol.*, 199(1):1–6. DOI: 10.1016/j.surfcoat.2005.03.031. 45

[28] Liang, W. 2003. Surface modification of AISI 304 austenitic stainless steel by plasma nitriding, *Appl. Surf. Sci.*, 211(1–4):308–314. https://doi.org/10.1016/S0169-4332(03)00260-5 DOI: 10.1016/s0169-4332(03)00260-5. 45

[29] Shah, M. S., Saleem, M., Ahmad, R., Zakaullah, M., Qayyum, A., and Murtaza, G. 2008. Langmuir probe characterization of nitrogen plasma for surface nitriding of AISI-4140 steel, *J. Mater. Process. Tech.*, 199(1–3):363–368. DOI: 10.1016/j.jmatprotec.2007.08.025. 45

[30] Li, C. X. 2010. Active screen plasma nitriding—an overview, *Surf. Eng.*, 26(1–2):135–141. DOI: 10.1179/174329409x439032. 45

[31] Dong, Y., Li, X., Tian, L., Bell, T., Sammons, R. L., and Dong, H. 2011. Towards long-lasting antibacterial stainless steel surfaces by combining double glow plasma silvering with active screen plasma nitriding, *Acta Biomater.*, 7(1):447–457. DOI: 10.1016/j.actbio.2010.08.009. 45

[32] Lin, Y.-C., Lin, C.-Y., and Chiu, P.-W. 2010. Controllable graphene N-doping with ammonia plasma, *Appl. Phys. Lett.*, 96(13):133110. DOI: 10.1063/1.3368697. 45

[33] Liu, X., Tian, W., Xu, W., Liang, W., and Xu, Z. 2007. Wear resistance of TiAl intermetallics by plasma alloying and plasma carburization, *Surf. Coat. Technol.*, 201(9–11):5278–5281. DOI: 10.1016/j.surfcoat.2006.07.144. 45

[34] Kaestner, P., Olfe, J., and Rie, K.-T. 2001. Plasma-assisted boriding of pure titanium and TiAl6V4, *Surf. Coat. Technol.*, pp. 142–144, pp. 248–252. DOI: 10.1016/s0257-8972(01)01244-0. 45

[35] Habibolahzadeh, A. and Haftlang, F. 2017. Duplex surface treatment of AISI 1045 steel via pack boriding and plasma nitriding: Characterization and tribological studies, *J. Tribol.*, 140(2):021602. DOI: 10.1115/1.4037465. 45

[36] Bazaka, K., Baranov, O., Cvelbar, U., Podgornik, B., Wang, Y., Huang, S., Xu, L., Lim, J. W. M., Levchenko, I., and Xu, S. 2018. Oxygen plasmas: A sharp chisel and handy trowel for nanofabrication, *Nanoscale*, 10:17494. DOI: 10.1039/c8nr06502k. 45, 47

[37] Xu, T., Yang, J., Liu, J., and Fu, Q. 2007. Surface modification of multiwalled carbon nanotubes by O2 plasma, *Appl. Surf. Sci.*, 253(22):8945–8951. DOI: 10.1016/j.apsusc.2007.05.028. 45

[38] Yang, X., Kimmel, Y. C., Fu, J., Koel, B. E., and Chen, J. G. 2012. Activation of tungsten carbide catalysts by use of an oxygen plasma pretreatment, *ACS Catal.*, 2(5):765–769. DOI: 10.1021/cs300081t. 46

[39] Yoshida, S., Hagiwara, K., Hasebe, T., and Hotta, A. 2013. Surface modification of polymers by plasma treatments for the enhancement of biocompatibility and controlled drug release, *Surf. Coat. Technol.*, 233:99–107. DOI: 10.1016/j.surfcoat.2013.02.042. 46

[40] Bazaka, K., Jacob, M. V., Crawford, R. J., and Ivanova, E. P. 2011. Plasma-assisted surface modification of organic biopolymers to prevent bacterial attachment, *Acta Biomater.*, 7(5):2015–2028. DOI: 10.1016/j.actbio.2010.12.024. 46

[41] Gupta, B., Hilborn, J., Hollenstein, C., Plummer, C. J. G., Houriet, R., and Xanthopoulos, N. 2000. Surface modification of polyester films by RF plasma, *J. Appl. Polym. Sci.*, 78(5):1083–1091. DOI: 10.1002/1097-4628(20001031)78:5%3C1083::aid-app170%3E3.0.co;2-5. 46

[42] Bazaka, K. and Jacob, M. V. 2009. Synthesis of radio frequency plasma polymerized nonsynthetic Terpinen-4-ol thin films, *Mater. Lett.*, 63(18–19):1594–1597. DOI: 10.1016/j.matlet.2009.04.025. 46

[43] Bazaka, K., Jacob, M. V., Truong, V. K., Wang, F., Pushpamali, W. A. A., Wang, J. Y. et al. 2010. Plasma-enhanced synthesis of bioactive polymeric coatings from monoterpene alcohols: a combined experimental and theoretical study, *Biomacromolecules*, 11(8):2016–2026. DOI: 10.1021/bm100369n. 47

[44] Bazaka, K., Jacob, M. V., Truong, V. K., Crawford, R. J., and Ivanova, E. P. 2011. The effect of polyterpenol thin film surfaces on bacterial viability and adhesion, *Polymers*, 3(1):388–404. DOI: 10.3390/polym3010388. 47

[45] Bazaka, K., Ketheesan, N., and Jacob, M. V. 2014. Polymer encapsulation of magnesium to control biodegradability and biocompatibility, *J. Nanosci. Nanotech.*, 14(10):8087–8093. DOI: 10.1166/jnn.2014.9409. 47

[46] Bazaka, K., Jacob, M. V., Taguchi, D., Manaka, T., and Iwamoto, M. 2011. Investigation of interfacial charging and discharging in double-layer pentacene-based metal-insulatormetal device with polyterpenol blocking layer using electric field induced, *Chem. Phys. Lett.*, 503(1–3):105–111. DOI: 10.1016/j.cplett.2010.12.072. 47

[47] Jacob, M. V., Bazaka, K., Taguchi, D., Manaka, T., and Iwamoto, M. 2012. Electron-blocking hole-transport polyterpenol thin films, *Chem. Phys. Lett.*, 528:26–28. DOI: 10.1016/j.cplett.2012.01.031. 47

[48] Bazaka, K., Destefani, R., and Jacob, M. V. 2016. Plant-derived cis-β-ocimene as a precursor for biocompatible, transparent, thermally-stable dielectric and encapsulating layers for organic electronics, *Sci. Rep.*, 6:38571. DOI: 10.1038/srep38571. 47

[49] Bazaka, K., Ahmad, J., Oelgemöller, M., Uddin, A., and Jacob, M. V. 2017. Photostability of plasma polymerized γ-terpinene thin films for encapsulation of OPV, *Sci. Rep.*, 7:45599. DOI: 10.1038/srep45599. 47

[50] Bazaka, K., Jacob, M. V., and Ostrikov, K. 2015. Sustainable life cycles of natural-precursor-derived nanocarbons, *Chem. Rev.*, 116(1):163–214. DOI: 10.1021/acs.chemrev.5b00566. 47

[51] Baranov, O., Levchenko, I., Bell, J., Lim, M., Huang, S., Xu, L., Wang, B., Aussems, D. U. B., Xu, S., and Bazaka, K. 2018. From nanometre to millimetre: A range of capabilities for plasma-enabled surface functionalization and nanostructuring, *Mater. Horiz.*, 5:765–798. DOI: 10.1039/c8mh00326b. 47

[52] Alancherry, S., Jacob, M. V., Prasad, K., Joseph, J., Bazaka, O., Neupane, R., Varghese, O. K., Baranov, O., Xu, S., Levchenko, I., and Bazaka, K. 2020. Tuning and fine morphology control of natural resource-derived vertical graphene, *Carbon*, 159:668–685. DOI: 10.1016/j.carbon.2019.10.060. 47

[53] Santhosh, N. M., Filipič, G., Tatarova, E., Baranov, O., Kondo, H., Sekine, M., Hori, M., Ostrikov, K., and Cvelbar, U. 2018. Oriented carbon nanostructures by plasma processing: Recent advances and future challenges, *Micromachines*, 9(11):565. DOI: 10.3390/mi9110565. 47

[54] Kurita, S., Yoshimura, A., Kawamoto, H., Uchida, T., Kojima, K., Tachibana, M., Molina-Morales, P., and Nakai, H. 2005. Raman spectra of carbon nanowalls grown by plasma-enhanced chemical vapor deposition, *J. Appl. Phys.*, 97:104320. DOI: 10.1063/1.1900297. 47

[55] Jiang, N., Wang, H. X., Zhang, H., Sasaoka, H., and Nishimura, K. 2010. Characterization and surface modification of carbon nanowalls, *J. Mater. Chem.*, 20:5070–5073. DOI: 10.1039/c0jm00446d. 47

[56] Shiji, K., Hiramatsu, M., Enomoto, A., Nakamura, M., Amano, H., and Hori, M. 2005. Vertical growth of carbon nanowalls using RF plasma-enhanced chemical vapor deposition, *Diamond Relat. Mater.*, 14:831–834. DOI: 10.1016/j.diamond.2004.10.021. 47

[57] Cott, D. J., Verheijen, M., Richard, O., Radu, I., De Gendt, S., van Elshocht, S., and Vereecken, P. M. 2013. Synthesis of large area carbon nanosheets for energy storage applications, *Carbon*, 88:59–65. DOI: 10.1016/j.carbon.2013.02.030. 47

[58] Wang, J., Zhu, M., Outlaw, R. A., Zhao, X., Manos, D. M., and Holloway, B. C. 2004. Synthesis of carbon nanosheets by inductively coupled radio-frequency plasma enhanced chemical vapor deposition, *Carbon*, 42:2867–2872. DOI: 10.1016/j.carbon.2004.06.035. 47

[59] Song, X., Liu, J., Yu, L., Yang, J., Fang, L., Shi, H., Du, C., Wei, D. 2014. Direct versatile PECVD growth of graphene nanowalls on multiple substrates, *Mater. Lett.*, 137:25–28. DOI: 10.1016/j.matlet.2014.08.125. 47

[60] Jiang, L., Yang, T., Liu, F., Dong, J., Yao, Z., Shen, C., Deng, S., Xu, N., Liu, Y., and Gao, H.-J. 2013. Controlled synthesis of large-scale, uniform, vertically standing graphene for high-performance field emitters, *Adv. Mater.*, 25:250–255. DOI: 10.1002/adma.201203902. 47

[61] Ghosh, S., Ganesan, K., Polaki, S. R., Ravindran, T. R., Krishna, N. G., Kamruddina, M., and Tyagia, A. K. 2014. Evolution and defect analysis of vertical graphene nanosheets, *J. Raman Spectrosc.*, 45(8):642–649. DOI: 10.1002/jrs.4530. 47

[62] Baranov, O., Bazaka, K., Xu, S., Lim, J. W. M., Cvelbar, U., and Levchenko, I. 2018. Formation of vertically oriented graphenes: What are the key drivers of growth?, *2D Mater.*, 5:044002. DOI: 10.1088/2053-1583/aad2bc. 47

[63] Prasad, K., Bandara, C., Kumar, S., Singh, G., Brockhoff, B., and Bazaka, K. 2017. Effect of precursor on antifouling efficacy of vertically-oriented graphene nanosheets, *Nanomaterials*, 7(7):E170. DOI: 10.3390/nano7070170. 47

[64] Jacob, M. V., Rawat, R. S., Ouyang, B., Bazaka, K., Kumar, S., Taguchi, D. et al. 2015. Catalyst-free plasma enhanced growth of graphene from sustainable sources, *Nano Lett.*, 15(9):5702–5708. DOI: 10.1021/acs.nanolett.5b01363. 47

[65] Jacob, M. V., Taguchi, D., Iwamoto, M., Bazaka, K., and Rawat, R. S. 2017. Resistive switching in graphene-organic device: Charge transport properties of graphene-organic device through electric field induced optical second harmonic generation and charge modulation spectroscopy, *Carbon*, 112:111–116. DOI: 10.1016/j.carbon.2016.11.005. 47

[66] Arora, N. and Sharma, N. N. 2014. Arc discharge synthesis of carbon nanotubes: Comprehensive review, *Diam. Relat. Mater.*, 50:135. DOI: 10.1016/j.diamond.2014.10.001. 47

[67] Ikuno, T., Katayama, M., Kamada, K., Hiwatashi, S., Ohkura, S., Honda, S., and Oura, K. 2003. Influence of plasma state on the structural property of vertically oriented carbon nanotubes grown by RF plasma-enhanced chemical vapor deposition, *Jpn. J. Appl. Phys.*, 42:6717. DOI: 10.1143/jjap.42.6717. 47

[68] Yen, J. H., Leu, I. C., Lin, C. C., and Hon, M. H. 2005. Synthesis of well-aligned carbon nanotubes by inductively coupled plasma chemical vapor deposition, *Appl. Phys. A*, 80(2)415–421. DOI: 10.1007/s00339-004-2545-x. 47

[69] Guo, B., Košiček, M., Fu, J., Qu, Y., Lin, G., Baranov, O., Zavašnik, J., Cheng, Q., Ostrikov, K., and Cvelbar, U. 2019. Single-crystalline metal oxide nanostructures synthesized by plasma enhanced thermal oxidation, *Nanomaterials*, 9(10):1405. DOI: 10.3390/nano9101405. 47

[70] Belmonte, T., Arnoult, G., Henrion, G., and Gries, T. 2011. Nanoscience with nonequilibrium plasmas at atmospheric pressure, *J. Phys. D, Appl. Phys.*, 44:363001. DOI: 10.1088/0022-3727/44/36/363001. 47

[71] Filipič, G., Baranov, O., Mozetič, M., and Cvelbar, U. 2015. Growth dynamics of copper oxide nanowires in plasma at low pressures, *J. Appl. Phys.*, 117:043304. DOI: 10.1063/1.4906501. 47

[72] Baranov, O., Filipič, G., and Cvelbar, U. 2019. Towards a highly-controllable synthesis of copper oxide nanowires in radio-frequency reactive plasma: Fast saturation at the targeted size, *Plasma Sour. Sci. Technol.*, 28:084002. DOI: 10.1088/1361-6595/aae12e. 47

[73] Guo, Y., Brian Ong, T. M., Levchenko, I., and Xu, S. 2018. Inductively and capacitively coupled plasmas at interface: A comparative study towards highly efficient amorphous-crystalline Si solar cells, *Appl. Surf. Sci.*, 427:486. DOI: 10.1016/j.apsusc.2017.08.125. 47

[74] Fang, M., Zhang, H., Sang, L., Cao, H., Yang, L., Ostrikov, K., Levchenko, I., and Chen, Q. 2017. Plasma-assisted ALD to functionalize PET: Towards new generation flexible gadgets, *Flex. Print. Electron.*, 2:022001. DOI: 10.1088/2058-8585/aa6add. 49

[75] Park, M., Oh, S., Kim, H., Jung, D., Choi, D., and Park, J. S. 2013. Gas diffusion barrier characteristics of Al_2O_3/Alucone films formed using trimethylaluminum, water and ethylene glycol for organic light emitting diode encapsulation, *Thin Solid Films*, 546:153–156. DOI: 10.1016/j.tsf.2013.05.017. 49

[76] Moosheimer, U. and Bichler, C. 1999. Plasma pretreatment of polymer films as a key issue for high barrier food packagings, *Surf. Coat. Tech.*, pp. 116–119, pp. 812–819. DOI: 10.1016/s0257-8972(99)00137-1. 49

[77] Wang, B. B., Zheng, K., Gao, D., Levchenko, I., Ostrikov, K., Keidar, M., and Zou, S. S. 2016. Plasma-chemical synthesis, structure and photoluminescence properties of hybrid graphene nanoflake-BNCO nanowall systems, *J. Mater. Chem. C*, 4:9788–9797. DOI: 10.1039/c6tc03871a. 49

[78] Wang, B. B., Qu, X. L., Zhu, M. K., Levchenko, I., Keidar, M., Zhong, X. X., Xu, S., and Ostrikov, K. 2017. Carbon nano flake-nanoparticle interface: A comparative study on structure and photolumine scent properties of carbon nanoflakes synthesized on nanostructured gold and carbon by hot filament CVD, *Carbon*, 124:391–402. DOI: 10.1016/j.carbon.2017.08.078. 49

[79] Xu, S., Huang, S. Y., Levchenko, I., Zhou, H. P., Wei, D. Y., Xiao, S. Q., Xu, L. X., Yan, W. S., and Ostrikov, K. 2011. Highly efficient silicon nanoarray solar cells by a single-step plasma-based process, *Adv. Energy Mater.*, 1:373–376. DOI: 10.1002/aenm.201100085. 49

[80] Primč, G., Levchenko, I., Kumar, S., Cvelbar, U., Mozetic, M., and Ostrikov, K. 2014. Imaging of the asymmetric DC discharge: Visualization to adjust plasma in the novel PECVD reactor, *IEEE Trans. Plasma Sci.*, 42(10):2564–2565. DOI: 10.1109/tps.2014.2315789. 50

[81] Zhou, H. P., Ye, X., Huang, W., Wu, M. Q., Mao, L. N., Yu, B., Xu, S., Levchenko, I., and Bazaka, K. 2019. Wearable, flexible, disposable plasma-reduced graphene oxide stress sensors for monitoring activities in austere environments, *ACS Appl. Mater. Interf.*, 11:15122–15132. DOI: 10.1021/acsami.8b22673. 51

[82] Levchenko, I., Keidar, M., Xu, S., Kersten, H., and Ostrikov, K. 2013. Low-temperature plasmas in carbon nanostructure synthesis, *J. Vac. Sci. Technol. B*, 31(5). DOI: 10.1116/1.4821635. 53

[83] Tsakadze, Z. L., Levchenko, I., Ostrikov, K., and Xu, S. 2007. Plasma-assisted self-organized growth of uniform carbon nanocone arrays, *Carbon*, 45(10):2022–2030. DOI: 10.1016/j.carbon.2007.05.030. 55

[84] Fang, J., Levchenko, I., and Ostrikov, K. 2014. Free-standing alumina nanobottles and nanotubes pre-integrated into nanoporous alumina membranes, *Sci. Technol. Adv. Mater.*, 15:045004. DOI: 10.1088/1468-6996/15/4/045004. 55

[85] Fang, J., Levchenko, I., van der Laan, T., Kumar, S., and Ostrikov, K. 2014. Multipurpose nanoporous alumina-carbon nanowall bi-dimensional nano-hybrid platform via catalyzed and catalyst-free plasma CVD, *Carbon*, 78:627–632. DOI: 10.1016/j.carbon.2014.07.053. 55

[86] Fang, J., Aharonovich, I., Levchenko, I., Ostrikov, K., Spizzirri, P. G., Rubanov, S., and Prawer, S. 2012. Plasma-enabled growth of single-crystalline SiC/AlSiC core-shell nanowires on porous alumina templates, *Cryst. Growth. Des.*, 12:2917–2922. DOI: 10.1021/cg300103a. 56

[87] Uehara, H., Kakiage, M., Sekiya, M., Sakuma, D., Yamonobe, T., Takano, N., Barraud, A., Meurville, E., and Ryser, P. 2009. Size-selective diffusion in nanoporous but flexible membranes for glucose sensors, *ACS Nano*, 3:924–932. DOI: 10.1021/nn8008728. 56

[88] Kipke, S. and Schmid, G. 2004. Nanoporous alumina membranes as diffusion controlling systems, *Adv. Funct. Mater.*, 14:1184–1188. DOI: 10.1002/adfm.200400193. 56

[89] Boss, C., Meurville, E., Sallese, J.-M., and Ryser, P. 2012. Size-selective diffusion in nanoporous alumina membranes for a glucose affinity sensor, *J. Membr. Sci.*, pp. 401–402, pp. 217–221. DOI: 10.1016/j.memsci.2012.02.004. 56

[90] Ma, X., Peng, S., Li, W., Liu, H., and Chen, Y. 2018. Efficient removal of low concentration methyl mercaptan by HKUST-1 membrane constructed on porous alumina granules, *CrystEngComm*, 20:407–411. DOI: 10.1039/c7ce01922j. 56

[91] Bertó-Roselló, F., Xifré-Pérez, E., Ferré-Borrull, J., and Marsal, L. F. 2018. 3D-FDTD modelling of optical biosensing based on gold-coated nanoporous anodic alumina, *Results Phys.*, 11:1008–1014. DOI: 10.1016/j.rinp.2018.10.067. 56

[92] Levchenko, I., Cvelbar, U., and Keidar, M. 2016. Graphene flakes in arc plasma: Conditions for the fast single-layer growth, *Graphene*, 5:81–89. DOI: 10.4236/graphene.2016.52009. 56

[93] Volotskova, O., Levchenko, I., Shashurin, A., Raitses, Y., Ostrikov, K., and Keidar, M. 2010. Single-step synthesis and magnetic separation of graphene and carbon nanotubes in arc discharge plasmas, *Nanoscale*, 2(10):2281. DOI: 10.1039/c0nr00416b. 59

[94] Pai, D. Z., Lacoste, D. A., and Laux, C. O. 2010. Nanosecond repetitively pulsed discharges in air at atmospheric pressure—the spark regime, *Plasma Sources Sci. Technol.*, 19:065015. DOI: 10.1088/0963-0252/19/6/065015. 61

[95] Chuang, A. T. H., Boskovic, B. O., and Robertson, J. 2006. Freestanding carbon nanowalls by microwave plasma-enhanced chemical vapour deposition, *Diam. Relat. Mater.*, 15(4–8):1103–1106. DOI: 10.1016/j.diamond.2005.11.004. 61

[96] Bundaleska, N., Tsyganov, D., Dias, A., Felizardo, E., Henriques, J., Dias, F. M., Abrashev, M., Kissovski, J., and Tatarova, E. 2018. Microwave plasma enabled synthesis of free standing carbon nanostructures at atmospheric pressure conditions, *Phys. Chem. Chem. Phys.*, 20:13810–13824. DOI: 10.1039/c8cp01896k. 61

CHAPTER 5

Perspectives and Trends

In addition to traditional processing of materials and material systems for, e.g., medicine, electronics and automotive industries where plasma-based tools are expected to play an increasingly important role as these fields transition to nanofabrication, the use of plasmas in the fields of cancer medicine, biotechnology, and biomass processing will also increase. Below we outline several rapidly-growing or emerging plasma fields that show great promise in addressing some of the most pressing challenges of the time.

Plasmas for the synthesis and assembly of complex nanostructured metamaterials.   The demand for complex nanostructured metamaterials and hierarchical nanostructures [1] will continue to rely on and drive the development of technological plasma systems that could support sophisticated, synthesis, and processing of material with control over their properties across macro-, micro-, and nano-scales. Currently, this level of precision and control in how fluxes of energy and mass are delivered to the surface generally comes at the expense of deposition rate, which hinders upscaling and translation into industry. To overcome this challenge, greater advantage should be taken of plasma-enabled self-organization through the design of innovative schemes and architectures that allow sophisticated control of highly energetic plasma jets and fluxes that could drive these processes [2].

Plasmas for medicine.   While the use of plasmas as a means to control surface properties of implants is well established, with examples ranging from enhancing surface wettability and promoting host cell-surface compatibility to protecting the surface from microbial colonization, corrosion and wear, the use of plasmas as a therapeutic tool is rapidly emerging. This is because there is a growing body of evidence that the unique combination of physical, chemical, and electromagnetic features produced in plasmas can be used to control the behavior of living cells and their systems. Based predominantly on the atmospheric pressure plasma jets and direct barrier discharges, these systems hold great promise for the treatment of resistant cancers, antibiotics-free tissue decontamination, stimulation of wound healing and tissue regeneration, and many other medical applications [3, 4]. Their future development and translation into clinics will depend on our ability to both gain deeper understanding of the individual and synergistic mechanisms and feedback loops that may underpin the observed biological responses [5], and then to develop systems that offer a sufficient level of control over the plasma-generated effects to ensure the needed level of therapeutic efficacy and selectivity.

Plasmas for food and water security.   With a looming threat of widespread shortages of food and water as the global population approaches 8 billion by 2024, means that can positively contribute to any stage of food cycle are becoming increasingly important. Decontamination of foods with plasma can significantly extend their shelf life, remove pesticide residue [6] and thus ensure greater food safety. Treatment of seed stocks before planting or after harvesting can positively contribute to crop productivity, by providing seeds with germination-, growth-, and fruiting-enhancing biochemical and biophysical stimuli, and extend their useful life in storage through decontamination [7, 8]. The treatment of industrial microorganisms can increase the efficiency of industrial fermentation [9]. Water, too, can be treated by plasmas to remove pollutants and pathogenic microorganisms [10], and enrich it with chemical species that can promote the growth and health of plants, fish, and microorganisms [11]. For these applications, scale, energy and cost efficiency, and compatibility with existing workflow and infrastructure are critical, thus considerable research efforts are devoted to solving these challenges [12, 13].

Plasmas for biowaste utilization, sustainable technology, and energy generation.   A topic closely related to food security is that of efficient bioresource utilization and resource conservation, which is an important step in closing the sustainable bioeconomy cycle [14]. The highly reactive chemistry and rapid heat generation inherent to plasmas may provide a suitable environment for controlled conversion of raw biomass and negative-value (waste) materials into useful chemicals [15], e.g., fuels with high energy density, feedstock for animals and industrial microorganisms, and precursors for chemical synthesis of value-added materials [16]. These plasmas can be driven directly by solar energy and use ambient air as the processing gas [17]. Plasmas can also contribute to sustainable energy generation by efficiently modifying the interfacial properties of nanomaterials used in photocatalysis, H_2 production, and photovoltaics [18].

Plasmas for space economy.   A growing interest and investment in the exploration and commercial exploitation of space and space resources will soon see an increasing number of space assets in orbit or permanently located on the surfaces of moons, planets, and asteroids. The useful lifetime of these assets is currently limited by the availability of suitable repair facilities that could be deployed along with the asset to provide ongoing maintenance to the asset. Plasma-based repair systems may be well suited to these challenging conditions, since the vacuum in space would obviate the need for most of the infrastructure associated with the use of low pressure plasma systems on Earth, e.g., chambers, pumps and other auxiliary equipment. This would allow plasmas to be generated using simples systems with an *open architecture* in an energy-efficient manner. It is in principle possible for plasma systems to become among the first fabrication tools used in permanent extra-terrestrial outposts [19], especially since conceptually similar plasma-based systems are already being extensively researched for propulsion applications [20, 21]. Plasma-based repair tools may be particularly useful given that plasma-made materials are considered promising for photonic thrust systems (ultra-light solar sails and light-reflecting coatings [22, 23]), advanced ultra-light propellant feed systems [24], novel materials with extra-high erosion resistance [25] and novel adaptive, self-healing, and self-regenerating materials for miniaturized space systems [26, 27].

5.1   REFERENCES

[1] Levchenko, I., Bazaka, K., Keidar, M., Xu, S., and Fang, J. 2018. Hierarchical multi-component inorganic metamaterials: Intrinsically driven self-assembly at nanoscale, *Adv. Mater.*, 30:1702226. DOI: 10.1002/adma.201702226. 73

[2] Adamovich, I., Baalrud, S. D., Bogaerts, A., Bruggeman, P. J., Cappelli, M., Colombo, V., Czarnetzki, U., Ebert, U., Eden, J. G., Favia, P. et al. 2017. The 2017 plasma roadmap: Low temperature plasma science and technology (topical review), *J. Phys. D, Appl. Phys.*, 50:323001. DOI: 10.1088/1361-6463/aa76f5. 73

[3] Chen, Z., Zhang, S., Levchenko, I., Beilis, I. I., and Keidar, M. 2017. In vitro demonstration of cancer inhibiting properties from stratified self-organized plasma-liquid interface, *Sci. Rep.*, 7(1):12163. DOI: 10.1038/s41598-017-12454-9. 73

[4] Bazaka, K., Bazaka, O., Levchenko, I., Xu, S., Ivanova, E. P., Keidar, M., and Ostrikov, K. 2017. Plasma-potentiated small molecules-possible alternative to antibiotics?, *Nano Futures*, 1(2):025002. DOI: 10.1088/2399-1984/aa80d3. 73

[5] Han, Z. J., Rider, A. E., Fisher, C., Laan, T. van der Kumar, S., Levchenko, I., and Ostrikov, K. 2014. Biological application of carbon nanotubes and graphene, *Carbon Nanotubes Graphene*, Elsevier, Chapter 12, pp. 279–312. DOI: 10.1016/b978-0-08-098232-8.00012-7. 73

[6] Zhou, R., Zhou, R., Yu, F., Xi, D., Wang, P., Li, J., Wang, X., Zhang, X., Bazaka, K., and Ostrikov, K. 2018. Removal of organophosphorus pesticide residues from lycium barbarum by gas phase surface discharge plasma, *Chem. Eng. J.*, 342:401–409. DOI: 10.1016/j.cej.2018.02.107. 74

[7] Niemira, B. A. 2012. Cold plasma decontamination of foods, *Ann. Rev. Food Sci. Technol.*, 3:125–142. DOI: 10.1146/annurev-food-022811-101132. 74

[8] Pankaj, S. K., Wan, Z., and Keener, K. M. 2018. Effects of cold plasma on food quality: A review, *Foods*, 7:4. DOI: 10.3390/foods7010004. 74

[9] Recek, N., Zhou, R., Zhou, R., Te'o, V. S. J., Speight, R. E., Mozetič, M., Vesel, A., Cvelbar, U., Bazaka, K., and Ostrikov, K. 2018. Improved fermentation efficiency of s. cerevisiae by changing glycolytic metabolic pathways with plasma agitation, *Sci. Rep.*, 8:8252. DOI: 10.1038/s41598-018-26227-5. 74

[10] Zhou, R., Zhou, R., Prasad, K., Fang, Z., Speight, R., Bazaka, K., and Ostrikov, K. 2018. Cold atmospheric plasma activated water as a prospective disinfectant: The crucial role of peroxynitrite, *Green Chem.*, 20:5276–5284. DOI: 10.1039/c8gc02800a. 74

[11] Zhou, R., Zhou, R., Wang, P., Luan, B., Zhang, X., Fang, Z., Xian, Y., Lu, X., Ostrikov, K., and Bazaka, K. 2019. Microplasma bubbles: Reactive vehicles for biofilm dispersal ACS, *Appl. Mater. Interf.*, 11:20660–20669. DOI: 10.1021/acsami.9b03961. 74

[12] Perinban, S., Orsat, V., and Raghavan, V. 2019. Nonthermal plasma-liquid interactions in food processing: A review, *Compr. Rev. Food Sci. Food Safety*, 18:1985–2008. DOI: 10.1111/1541-4337.12503. 74

[13] Muhammad, A. I., Liao, X., Cullen, P. J., Liu, D., Xiang, Q., Wang, J., Chen, S., Ye, X., and Ding, T. 2018. Effects of nonthermal plasma technologyon functional food components, *Compr. Rev. Food Sci. Food Safety*, 17:1379–1394. DOI: 10.1111/1541-4337.12379. 74

[14] Tamilselvi, R., Ramesh, M., Lekshmi, G. S., Bazaka, O., Levchenko, I., Bazaka, K., and Mandhakini, M. 2020. Graphene oxide-based supercapacitors from agricultural wastes: A step to mass production of highly efficient electrodes for electrical transportation systems, *Renew. Energy*, 151:731–739. DOI: 10.1016/j.renene.2019.11.072. 74

[15] Zhou, Renwu, Zhou, Rusen, Zhang, X., Fang, Z., Wang, X., Speight, R., Wang, H., Doherty, W., Cullen, P. J., Ostrikov, K., and Bazaka, K. 2019. High-performance plasma-enabled biorefining of microalgae to value-added products, *ChemSusChem*, 12:4976–4985. DOI: 10.1002/cssc.201901772. 74

[16] Watson, J., Zhang, Y., Si, B., Chen, W. T., and Souza, R. 2018. Gasification of biowaste: A critical review and outlooks, *Renew. Sustain. Energy Rev.*, 83:1–17. DOI: 10.1016/j.rser.2017.10.003. 74

[17] Zhou, R., Zhou, R., Xian, Y., Fang, Z., Lu, X., Bazaka, K., Bogaerts, A., and Ostrikov, K. 2020. Plasma-enabled catalyst-free conversion of ethanol to hydrogen gas and carbon dots near room temperature, *Chem. Eng. J.*, 382:122745. DOI: 10.1016/j.cej.2019.122745. 74

[18] Xiang, Q., Ma, X., Zhang, D., Zhou, H., Liao, Y., Zhang, H., Xu, S., Levchenko, I., and Bazaka, K. 2019. Interfacial modification of titanium dioxide to enhance photocatalytic efficiency towards H_2 production, *J. Colloid Interf. Sci.*, 556:376–385. DOI: 10.1016/j.jcis.2019.08.033. 74

[19] Levchenko, I., Xu, S., Mazouffre, S., Keidar, M., and Bazaka, K. 2018. Mars colonization: Beyond getting there, *Glob. Chall.*, 2:1800062. DOI: 10.1002/gch2.201800062. 74

[20] Levchenko, I., Bazaka, K., Ding, Y. et al. 2018. Space micropropulsion systems for cube-sats and small satellites: From proximate targets to furthermost frontiers, *Appl. Phys. Rev.*, 5:011104. DOI: 10.1063/1.5007734. 74

[21] Levchenko, I., Bazaka, K., Belmonte, T., Keidar, M., and Xu, S. 2018. Advanced materials for next generation spacecraft, *Adv. Mater.*, 30:1802201. DOI: 10.1002/adma.201802201. 74

[22] Levchenko, I., Bazaka, K., Mazouffre, S., and Xu, S. 2018. Prospects and physical mechanisms for photonic space propulsion, *Nature Photon.*, 12:649–657. DOI: 10.1038/s41566-018-0280-7. 74

[23] Levchenko, I., Xu, S., Teel, G., Mariotti, D., Walker, M. L. R., and Keidar, M. 2018. Recent progress and perspectives of space electric propulsion systems based on smart nano-materials, *Nat. Commun.*, 9:879. DOI: 10.1038/s41467-017-02269-7. 74

[24] Singhal, N., Levchenko, I., Huang, S., Xu, L., Potrivitu, G. C., Cherkun, O., Fang, J., Bazaka, K., and Xu, S. 2019. 3D-printed multilayered reinforced material system for gas supply in CubeSats and small satellites, *Adv. Eng. Mater.*, 21:1900401. DOI: 10.1002/adem.201900401. 74

[25] Ding, Y., Wang, L., Fan, H., Li, H., Xu, W., Wei, L., Li, P., and Yu, D. 2019. Simulation research on magnetic pole erosion of Hall thrusters, *Phys. Plasmas*, 26:023520. DOI: 10.1063/1.5077041. 75

[26] Levchenko, I., Keidar, M., Cantrell, J., Wu, Y. L., Kuninaka, H., Bazaka, K., and Xu, S. 2018. Explore space using swarms of tiny satellites, *Nature*, 56:185–187. DOI: 10.1038/d41586-018-06957-2. 75

[27] Levchenko, I., Xu, S., Wu, Y., and Bazaka, K. 2020. Hopes and concerns for astronomy of satellite constellations. *Nat. Astron.*, (in press) DOI: 10.1038/s41550-020-1141-0 75

CHAPTER 6

Conclusion

In this mini-book, we have briefly reviewed a number of representative systems and architectures that are currently used for plasma-based material fabrication and processing, introducing the reader to the general schematics and key control mechanisms characteristic of each of these architectures. We also outlined the principal parameters of plasmas for these systems, and their practical implications for material synthesis, assembly, and modification, briefly touching on the mathematical description of the plasma behavior in these technological setups. While primarily focusing on established technologies such material etching, deposition, and modification, we briefly discussed the main challenges for their application in nanofabrication and in other emerging fields, including in large-scale biomass valorization, cancer medicine, and space technology.

Authors' Biographies

OLEG O. BARANOV

Oleg O. Baranov received his Ph.D. and D.Sc. degrees in aircraft technology and materials science from National Aerospace University, Kharkiv, Ukraine, in 2000 and 2016, respectively. Since 2000, he has been a Faculty Member with the National Aerospace University focusing on the development and optimization of aircraft engines. His current research interests include plasma physics and nanotechnology, including PVD coating, magnetron and vacuum-arc deposition, plasma control and diagnostics in deposition setups, mechanical properties of materials and thin films, and surface plasma processing.

IGOR LEVCHENKO

Igor Levchenko was Senior Research Fellow at the Commonwealth Scientific Industrial Research Organization, Sydney, NSW, Australia. He is currently a Research Scientist with the Plasma Sources and Applications Centre/Space Propulsion Centre, Nanyang Technological University, Singapore, and an Adjunct Professor with the School of Chemistry, Physics, and Mechanical Engineering, Queensland University of Technology, Brisbane, QLD, Australia. He has authored or co-authored more than 200 journal articles, as well as several books and book chapters. His current research interests include the development of plasma-enabled strategies for synthesis and functionalization of hierarchical multicomponent metamaterials, their application in energy harvesting and storage, electronics and medicine, and electric propulsion systems for space applications.

SHUYAN XU

Shuyan Xu is a Professor of plasma physics with a combination of experimental, analytical, and computational knowledge and experience. He conceived and founded the Plasma Sources and Applications Centre, Nanyang Technological University, Singapore, that currently comprises several research units, namely Plasma Processing Laboratory, Advanced Materials and Nanostructures Laboratory, Photovoltaic Research Laboratory, and Space Propulsion Centre. He has authored or co-authored approximately 230 journal publications, 2 books, and several book chapters, delivered close to 70 keynote addresses and invited talks, and holds 10 international patents. His current research interests include physics and applications of high-density plasmas, rotating magnetic field current drive in fusion plasma, space electric propulsion, development of plasma facilities and diagnostics, and application of nonequilibrium plasma for silicon photovoltaics, as well as plasma-aided nanofabrication and processing.

KATERYNA BAZAKA

Kateryna Bazaka is an Associate Professor with the College of Engineering and Computer Science and Associate Director (Education), Research School of Electrical Energy and Materials Engineering, Australian National University. Her current research interests include the development and application of low-temperature plasmas for the synthesis and processing of functional nanostructured materials for electronics, energy, and biomedical applications, sustainable and space technologies, and direct and solution-mediated treatment of microorganisms and cancer cells.

Printed in the United States
by Baker & Taylor Publisher Services